GOVERNING CLIMATE CHANGE

Cities are no longer just places to live in. They are significant actors on the global stage, and nowhere is this trend more prominent than in the world of transnational climate change governance (TCCG). Through transnational networks that form links between cities, states, international organizations, corporations, and civil society, cities are developing and implementing norms, practices, and voluntary standards across national boundaries. In introducing cities as transnational lawmakers, Jolene Lin provides an exciting new perspective on climate change law and policy, offering novel insights about the reconfiguration of the state and the nature of international lawmaking as the involvement of cities in TCCG blurs the public/private divide and the traditional strictures of 'domestic' versus 'international'. This illuminating book should be read by anyone interested in understanding how cities – in many cases, more than the countries in which they are located – are addressing the causes and consequences of climate change.

Jolene Lin is Associate Professor of Law at the National University of Singapore and Director of the Asia Pacific Centre for Environmental Law. She has published widely in leading international journals such as the *European Journal of International Law, Legal Studies*, and *Journal of Environmental Law*. Jolene Lin is a member of the editorial boards of the *Journal of Environmental Law, Climate Law*, and the *Chinese Journal of Environmental Law*, as well as a member of the IUCN Academy of Environmental Law Research Committee.

CAMBRIDGE STUDIES ON ENVIRONMENT, ENERGY AND NATURAL RESOURCES GOVERNANCE

Cambridge Studies on Environment, Energy and Natural Resources Governance publishes foundational monographs of general interest to scholars and practitioners within the broadly defined fields of sustainable development policy, including studies on law, economics, politics, history, and policy. These fields currently attract unprecedented interest due to both the urgency of developing policies to address climate change, the energy transition, food security and water availability and, more generally, to the progressive realization of the impact of humans as a geological driver of the state of the Earth, now called the 'Anthropocene'.

The general editor of the series is Professor Jorge E. Viñuales, the Harold Samuel Chair of Law and Environmental Policy at the University of Cambridge and the Founder and First Director of the Cambridge Centre for Environment, Energy and Natural Resource Governance (C-EENRG).

Governing Climate Change

GLOBAL CITIES AND TRANSNATIONAL LAWMAKING

JOLENE LIN

Faculty of Law, National University of Singapore

CAMBRIDGE
UNIVERSITY PRESS

CAMBRIDGE
UNIVERSITY PRESS

University Printing House, Cambridge CB2 8BS, United Kingdom

One Liberty Plaza, 20th Floor, New York, NY 10006, USA

477 Williamstown Road, Port Melbourne, VIC 3207, Australia

314–321, 3rd Floor, Plot 3, Splendor Forum, Jasola District Centre,
New Delhi – 110025, India

79 Anson Road, #06–04/06, Singapore 079906

Cambridge University Press is part of the University of Cambridge.

It furthers the University's mission by disseminating knowledge in the pursuit of education, learning, and research at the highest international levels of excellence.

www.cambridge.org
Information on this title: www.cambridge.org/9781108424851
DOI: 10.1017/9781108347907

First published 2018

Printed in the United States of America by Sheridan Books, Inc.

A catalogue record for this publication is available from the British Library.

Library of Congress Cataloging-in-Publication Data
Names: Lin, Jolene.
Title: Governing climate change : global cities and transnational lawmaking / Jolene Lin.
Description: New York : Cambridge University Press, 2018. | Series: Cambridge studies on environment, energy and natural resources governance
Identifiers: LCCN 2018011575 | ISBN 9781108424851 (hardback)
Subjects: LCSH: Climate change mitigation – International cooperation. | Climatic changes – Law and legislation. | Municipal corporations. | Global warming – Law and legislation. | Global warming – International cooperation. | Climatic changes – International cooperation. | Environmental law, International. | BISAC: LAW / Environmental.
Classification: LCC K3585.5 .L56 2018 | DDC 344.04/633–dc23
LC record available at https://lccn.loc.gov/2018011575

ISBN 978-1-108-42485-1 Hardback
ISBN 978-1-108-44098-1 Paperback

Contents

Acknowledgements

I owe a great debt of gratitude to many people for their support and contribution towards the completion of this book. *Governing Climate Change* is a revised version of my PhD thesis, which I defended at Erasmus University Rotterdam in October 2017. Special thanks go to my promoter, Professor Dr Ellen Hey, for her insightful advice and many helpful conversations over dinner and on Skype. At the University of Hong Kong, where I started my academic career, I am deeply grateful to Professor Johannes Chan, former dean of the law school, for his support and encouragement. It would not have been possible to complete the thesis and this book otherwise. To Thomas Cheng, Marco Wan, and Ernest Lim, thank you for the friendship, laughter, and intellectual (as well as non-intellectual) debates. To Mi Zhou, thank you for your company in the office during the evenings and weekends, Ethiopian coffee, and wicked humour. At Erasmus University Rotterdam, I thank Jing Hiah, Anna Sting, and Ryan Gauthier for their friendship. In Utrecht, thanks to Evelyn Wan for inviting me to join a lovely writing group. To Diane Lek, I am grateful for your friendship and encouragement all these years.

I also thank Joanne Scott for always making time, despite an incredibly busy schedule, to impart words of wisdom and encouragement, as well as for painstakingly reading many drafts of this book.

To my parents, thank you for your love and for working so hard to give us the education and life experiences that have allowed us to flourish. To my husband, Thomas R. Wind, thank you for your love and for standing by me in everything I do. Thank you for the laughter and being my oasis of calmness and positive thinking. To my sister, Joycelyn Lin, I am always grateful for your love and generosity. Finally, to my son, Lucas Kai Wind, who is four months old at the time of writing these acknowledgements, it was a joy to complete this book in anticipation of your arrival. During those nine months, we had adventures in many great cities – London, Beijing, Hong Kong, and Amsterdam – and spent countless hours in the library. I dedicate this book to you.

I have endeavoured to state the law as of 1 January 2017.

Abbreviations

C40 Cities Climate Leadership Group	C40
Carbon capture and storage	CCS
Carbon dioxide	CO_2
Carbon dioxide equivalent	CO_2e
Chlorofluorocarbons	CFCs
Clean Development Mechanism	CDM
Conference of the Parties	COP
Convention on Biological Diversity	CBD
European Union	EU
European Union Emissions Trading Scheme	EU ETS
Intergovernmental Panel on Climate Change	IPCC
ICLEI – Local Governments for Sustainability	ICLEI
International Emissions Trading Association	IETA
Measurement, reporting, and verification	MRV
United Cities and Local Governments	UCLG
United Nations Development Programme	UNDP
United Nations Environment Programme	UNEP
United Nations Framework Convention on Climate Change	UNFCCC
World Business Council for Sustainable Development	WBCSD
World Health Organization	WHO
World Trade Organization	WTO

1

Global Cities, Climate Change, and Transnational Lawmaking

1.1 INTRODUCTION

On 12 December 2015, when French Foreign Minister Laurent Fabius announced that a new climate change agreement had been signed, cheers erupted in the negotiation hall and elsewhere around the world.[1] States had finally concluded more than two decades of difficult multilateral negotiations.[2] However, there should be no illusions that we are on track to averting dangerous human interference with the climate system.[3] As noted 'with concern' in the Paris decision, based on the mitigation pledges that states submitted in advance of the Conference of the Parties (COP) in Paris, global greenhouse gas (GHG) emission levels will reach 55 gigatonnes in 2030.[4] This far exceeds the 40 gigatonne limit necessary to hold the increase in the global average temperature to below 2 °C above pre-industrial

[1] Fiona Harvey, 'Paris Climate Change Agreement: The World's Greatest Diplomatic Success' *Guardian* (14 December 2015).

[2] For discussion of the Paris Agreement, see e.g. Daniel Bodansky, 'The Legal Character of the Paris Agreement' (2016) 25(2) *Review of European Community and International Environmental Law* 142; Christina Voigt, 'The Compliance and Implementation Mechanism of the Paris Agreement' (2016) 25 (2) *Review of European Community and International Environmental Law* 161.

[3] Article 2 of the United Nations Convention on Climate Change 1771 UNTS 163 (UNFCCC) states that the ultimate objective of the treaty is the stabilization of greenhouse gas concentrations in the atmosphere at a level that would prevent dangerous anthropogenic interference with the climate system. Such a level should be achieved within a time frame sufficient to allow ecosystems to adapt naturally to climate change, to ensure that food production is not threatened, and to enable economic development to proceed in a sustainable manner. Yamin and Depledge argue that this objective is akin to an environmental quality standard; Farhana Yamin and Joanna Depledge, *The International Climate Change Regime: A Guide to Rules, Institutions and Procedures* (Cambridge University Press 2004), pg. 61.

[4] Para. 17, Decision 1/CP. 21: Adoption of the Paris Agreement, UNFCCC, Report of the Conference of the Parties on its twenty-first session, held in Paris from 30 November to 13 December 2015. The Paris Agreement entered into force on 4 November 2016, thirty days after the date on which at least fifty-five parties to the UNFCCC, accounting in total for at least an estimated 55 per cent of the total global GHG emissions, have deposited their instruments of ratification, acceptance, approval, or accession (Article 21(1)); UNFCCC, 'The Paris Agreement', online: http://unfccc.int/paris_agreement/items/9485.php (accessed on 1 December 2016).

levels.[5] While the Paris Agreement requires states to progressively ratchet up their climate mitigation targets,[6] it would not be wise to rely on states alone to address climate change. Solving a complex problem in a complex global society will require action beyond what states can shoulder. Tackling climate change requires pragmatic deliberation involving multiple sources of knowledge and experience, not simply the top-down involvement of 'increasingly detached and under-resourced diplomats paralyzed by geopolitical power plays, hidden value systems, or zero-sum distributional calculations'.[7]

While government delegates were in marathon negotiating sessions trying to conclude the Paris Agreement, banks, corporations, think tanks, consultancies, and various other organizations were holding 'side events' at multiple venues across Paris.[8] At the Climate Summit for Local Leaders, mayors from around the world gathered to discuss climate change and to highlight the significant role that cities play in reducing GHG emissions and increasing society's resilience to the impacts of climate change. At the end of the summit, the city leaders delivered a declaration intended to 'demonstrate their global leadership on climate policies'.[9] Mayors who signed the Paris City Hall Declaration undertook commitments to '[a]dvance and exceed the expected goals of the 2015 Paris Agreement' and 'deliver up to 3.7 gigatons of urban [GHG] emissions reductions annually by 2030 – the equivalent of up to 30 per cent of the difference between current national commitments and the 2 degree emissions reduction pathway identified by the scientific community'.[10] Following the Paris City Hall Declaration, the Lima-Paris Action Agenda (Focus on Cities) proposed a Five Year Vision to accelerate climate action in cities.[11] The aim is that, by the year 2020, 'local action and partnerships should be the new

[5] Ibid.

[6] Article 4(3) of the Paris Agreement states: 'Each Party's successive nationally determined contribution will *represent a progression beyond the Party's then current nationally determined contribution and reflect its highest possible ambition*, reflecting its common but differentiated responsibilities and respective capabilities, in the light of different national circumstances' (emphasis mine).

[7] Joost Pauwelyn, Ramses Wessel, and Jan Wouters, 'Informal International Lawmaking: An Assessment and Template to Keep It Both Effective and Accountable' in Joost Pauwelyn, Ramses Wessel, and Jan Wouters (eds), *Informal International Lawmaking* (Oxford University Press 2012), pg. 526.

[8] For details of the official side events coordinated by the UNFCCC secretariat as well as events independently organized by observer organizations relating to the climate change negotiation process, see UNFCCC, *Side Events and Exhibits One Stop Shop*, online: http://unfccc.int/parties_and_observers/ngo/items/9325.php#Side%2520events (accessed on 1 July 2016).

[9] Climate Summit for Local Leaders: Cities for Climate (4 December 2015), online: http://climatesummitlocalleaders.paris (accessed on 1 March 2016).

[10] *Paris City Hall Declaration: A Decisive Contribution to COP21*, online: http://climatesummitlocalleaders.paris/content/uploads/sites/16/2016/01/CLIMATE-SUMMIT-LOCAL-LEADERS-POLITICAL-DECLARATION-PARIS-DEC-4–2015.pdf (accessed on 1 March 2016).

[11] Lima-Paris Climate Agenda, 'Cities and Regions Launch Major Five-Year Vision to Take Action on Climate Change', online: http://newsroom.unfccc.int/lpaa/cities-subnationals/lpaa-focus-cities-regions-across-the-world-unite-to-launch-major-five-year-vision-to-take-action-on-climate-change/ (accessed on 1 March 2016).

norm globally'.[12] In an interview with the *Financial Times*, the mayor of Paris, Anne Hidalgo, said, '[Cities] are more practical; we have the capacity to act faster and the decisions are closer to reality. We can mobilize all actors, public and private … [which is] more complicated for the state'.[13]

Mayor Hidalgo's opinion captures the essence of how cities perceive and situate themselves within the contemporary global effort to govern climate change. Cities do not claim to be simply implementers of international climate policy; they have positioned themselves as central participants and stakeholders, in their own right, of the global climate governance effort.[14] While early efforts by cities to address climate change (1990s–early 2000s) were mainly concerned with driving local action in the face of national recalcitrance and stalemate in international negotiations, cities today aim to play a prominent role in global climate change governance, including the formal international lawmaking process – the United Nations Framework Convention on Climate Change (UNFCCC).[15] In 2010, local and sub-national governments were conferred recognition as 'governmental stakeholders' within the UNFCCC regime in the Cancun Agreements.[16]

The ambition to take on a global role may be viewed as a natural extension of the increasingly significant participation of cities in multi-level climate governance arrangements, particularly in the European Union (EU). Within the EU, cities such as Southampton (United Kingdom) and Munich (Germany) are leading in terms of their development of local climate change strategies as well as their active engagement in climate protection networks such as Climate Alliance and Energy Cities.[17] As authority within the EU has not only shifted upwards from member states to European institutions but has also dispersed downwards to sub-national levels, municipal networks cooperate with each other to increase their influence and

[12] Lima-Paris Climate Agenda, 'A 5-Year Vision', online: http://newsroom.unfccc.int/media/544092/lpaa-five-year-vision.pdf, pg. 4 (accessed on 1 March 2016).

[13] Michael Stothard, 'Mayors Call for More Powers to Fight Climate Change' *Financial Times* (4 December 2015).

[14] Michele Acuto, 'The New Climate Leaders?' (2013) 39 *Review of International Studies* 835.

[15] It should also be noted that earlier urban climate action efforts primarily focused on ways to link climate change to issues already on the local agenda, such as improving air quality. Since the 2000s, there have been shifts in cities' climate governance agendas towards the need to scale localized actions and impacts up to the global level in order to achieve aggregate global effects; Harriet Bulkeley, 'Cities and the Governing of Climate Change' (2010) 35 *Annual Review of Environment and Resources* 229.

[16] Decision 1/CP.16 The Cancun Agreements: Outcome of the Work of the Ad Hoc Working Group on Long-Term Cooperative Action under the Convention, para. 7.

[17] In terms of number of members, Climate Alliance claims to be the 'largest European city network dedicated to climate action'; Climate Alliance, 'About Us', online: www.climatealliance.org/about-us.html; Energy Cities is a European network of local authorities focused on energy transition and sustainable energy. Amongst its key objectives is 'to represent [local authorities'] interests and influence the policies and proposals made by European Union institutions in the fields of energy, environmental protection and urban policy'; Energy Cities, 'Main Objectives', online: www.energy-cities.eu/-Association,8- (accessed on 1 March 2016).

solicit EU funding.[18] At the same time, these networks can help the European Commission implement EU policies through the exchange of best practices and the production of standards which member cities are required to implement.[19] In brief, cities have evolved from being 'passive implementers' to 'active co-decision makers' in the EU climate governance context.[20] Their experience of working across various levels of governance and cooperating through networks to develop and implement governance initiatives has created fertile ground for 'up-scaling' these efforts to the transnational level.

As global governors – i.e. 'authorities who exercise power across borders for purposes of affecting policy'[21] – cities have created networks that connect thousands of cities across the globe. These networks operate across the public-private divide, forming partnerships and cooperating with other actors, be they multinational corporations, global non-governmental organizations (NGOs), or philanthropic foundations.[22] They seek to distil and disseminate authoritative and credible information to their member cities throughout the world. The networks also aggregate the influence of cities so that they have a more prominent collective voice in international forums such as the UNFCCC.

There are four transnational networks working in the area of city-focused and city-driven climate governance: (i) ICLEI – Local Governments for Sustainability (ICLEI), (ii) United Cities and Local Government (UCLG), (iii) the World Mayors Council on Climate Change, and (iv) the C40 Cities Climate Leadership Group (C40). Through these networks, cities around the world create physical and virtual platforms to share best practices and experience. They utilize information and communication technologies to create collective knowledge as well as enhance transparency, which in turn fosters legitimacy.[23] For ease of reference, this book will refer to these transnational networks of cities as *city networks*.

[18] Kristine Kern and Harriet Bulkeley, 'Cities, Europeanization and Multi-Level Governance: Governing Climate Change through Transnational Municipal Networks' (2009) 47 *Journal of Common Market Studies* 309, pg. 313.

[19] Ibid.

[20] Kristine Kern and Arthur P. J. Mol, 'Cities and Global Climate Governance: From Passive Implementers to Active Co-Decision-Makers' in Mary Kaldor and Joseph E. Stiglitz (eds), *The Quest for Security: Protection without Protectionism and the Challenge of Global Governance* (Columbia University Press 2013), pg. 288.

[21] Deborah D. Avant, Martha Finnemore, and Susan K. Sell (eds), *Who Governs the Globe?* (Cambridge University Press 2010), pg. 2.

[22] See discussion in Chapter 5.

[23] An example drawn from Chapter 5 is the carbon*n*® Climate Registry. This is an online reporting platform that allows sub-national governments to publicly report their climate actions. Anyone with an Internet connection can gain access to the carbon*n* Climate Registry to monitor whether a city has fulfilled its climate action commitments. Such transparency mechanisms allow the media, civil society, and citizens to play a quasi-monitoring and enforcement function. On its website, carbon*n* Climate Registry is described as 'designed as the global response of local and sub-national governments towards measurable, reportable and verifiable climate action'; online: http://carbonn.org (accessed on 1 July 2016).

Briefly, ICLEI coordinates local government representation in several UN processes related to Agenda 21 and the Habitat Agenda.[24] It has observer status at the UNFCCC and has been a leading advocate for greater recognition of the role of local and sub-national governments in the international climate change regime.[25] UCLG's stated mission is 'to be the united voice and world advocate of democratic local self-government', and it facilitates programmes and partnerships to build the capacity of local governments.[26] UCLG's global agenda includes disaster risk reduction, the 2030 Agenda for Sustainable Development, water and sanitation, and climate change.[27] The third network is the World Mayors Council on Climate Change, founded in December 2005 by the mayor of Kyoto (Japan) soon after the Kyoto Protocol entered into force in February 2005.[28] The network receives technical and strategic support from ICLEI.[29] Since the adoption of the 2012 Seoul Declaration of Local Governments on Energy and Climate Mitigation, the World Mayors Council on Climate Change has been relatively quiet and primarily involved in supporting other networks (e.g. ICLEI) and initiatives such as the Compact of Mayors (which will be discussed in detail later).[30] Finally, C40 can be described as being the most well-known network of cities addressing climate change. It has rapidly gained prominence because of its unique focus on global cities and climate change (while the other three networks address climate change as one of many issue areas that they work in) as well as the partnerships it has fostered with high-profile organizations such as the World Bank and the Clinton Foundation.[31] This book will focus on C40, because both its global city membership and modus operandi render it well suited for a study of how global cities engage in hybrid public-private governance arrangements in addressing climate change and how these arrangements produce norms, practices, and voluntary standards.

The United Nations (UN) has also embraced the urban agenda. For example, in support of the previously mentioned Five Year Vision, the UN has formed a 'sub-national action hub' that will entail a wide mobilization of UN agencies to help cities and regions increase the scale and number of climate actions and plans.[32]

24 ICLEI, 'Recognizing, Engaging and Resourcing Local Governments', online: www.iclei.org/activities/advocacy.html (accessed on 1 July 2016).

25 ICLEI at COP21, online: www.iclei.org/activities/advocacy/cop21.html (accessed on 1 July 2016).

26 United Cities and Local Government, online: www.uclg.org/en (accessed on 1 July 2016).

27 Ibid.

28 World Mayors Council on Climate Change, online: www.worldmayorscouncil.org/home.html (accessed on 1 July 2016).

29 Ibid.

30 2012 Seoul Declaration of Local Governments on Energy and Climate Mitigation, online: www.worldmayorscouncil.org/fileadmin/Documents/Seoul/2012_SeoulDeclaration_ofLocalGovernments_onEnergyandClimateMitigation.pdf (accessed on 1 July 2016).

31 See discussion in Chapter 5.

32 Lima-Paris Climate Agenda, 'Cities and Regions Launch Major Five-Year Vision to Take Action on Climate Change'. The UN agencies include the UN Environment Programme (UNEP), the UN Development Programme (UNDP), UN-Habitat, the World Health Organization (WHO), the World Bank, and the UN Industrial Development Organization (UNIDO).

In 2018, the Intergovernmental Panel on Climate Change (IPCC) will hold a major scientific conference to further develop scientific understanding of climate change and cities, a key recognition of the role of cities in addressing climate change.[33] The Sustainable Development Goals (SDGs) finalized by negotiators from UN member states in September 2015 also recognize the significance of the urban unit in determining the state of the environment. SDG 11 challenges policymakers and governments to '[m]ake cities and human settlements inclusive, safe, resilient and sustainable'.[34] Cities are also working directly with international organizations like the UN Human Settlements Programme (UN-Habitat) and the World Bank to address climate change as part of a larger, multipronged urban sustainable development agenda.[35] In doing so, cities are engaging in transnational relations that bypass national governments and forging a direct link between the local level and international organizations.

1.2 THE SUBJECT MATTER OF THIS BOOK: THE LAWMAKING ROLE OF FIVE CITIES IN TRANSNATIONAL CLIMATE CHANGE GOVERNANCE

Although there is a large body of literature on 'cities and climate change governance' that continues to grow rapidly, few scholars have considered the legal effect and normative relevance of cities' governance activities.[36] This book aims to fill this gap in the literature by examining the emergence of cities as actors that are producing and implementing norms, practices, and voluntary standards that transcend state boundaries to steer the behaviour of cities towards reducing GHG emissions and

[33] Cities and Climate Change Science Conference (Edmonton, Canada, 5–7 March 2018), online: www.citiesipcc.org (accessed on 1 August 2017).

[34] UN Sustainable Development Goals, online: https://sustainabledevelopment.un.org/?menu=1300 (accessed on 1 August 2016).

[35] For example, the Low Carbon, Livable Cities initiative sees the World Bank eschewing the traditional approach of multilateral cooperation and bypassing the state to work directly with city officials; see discussion in Section 3.3.

[36] On cities and climate change, see for example Michele M. Betsill and Harriet Bulkeley, 'Transnational Networks and Global Environmental Governance: The Cities for Climate Protection Program' (2004) 48 *International Studies Quarterly* 471; Harriet Bulkeley and Michele Betsill, 'Rethinking Sustainable Cities: Multilevel Governance and the "Urban" Politics of Climate Change' (2005) 14 *Environmental Politics* 42; Ute Collier, 'Local Authorities and Climate Protection in the European Union: Putting Subsidiarity into Practice?' (1997) 2 *Local Environment* 39; Gard Lindseth, 'The Cities for Climate Protection Campaign (CCPC) and the Framing of Local Climate Policy' (2004) 9 *Local Environment* 325; Renske den Exter, Jennifer Lenhart, and Kristine Kern, 'Governing Climate Change in Dutch Cities: Anchoring Local Climate Strategies in Organisation, Policy and Practical Implementation' (2014) *Local Environment: The International Journal of Justice and Sustainability*; Joyeeta Gupta, Ralph Lasage, and Tjeerd Stam, 'National Efforts to Enhance Local Climate Policy in the Netherlands' (2007) 4 *Environmental Sciences* 171; Heleen Lydeke, P. Mees, and Peter P. J. Driessen, 'Adaptation to Climate Change in Urban Areas: Climate-Greening London, Rotterdam, and Toronto' (2011) 2 *Climate Law* 251.

developing low-carbon alternatives for the future. These norms, practices, and voluntary standards impose limitations on how cities develop by requiring them to take climate risks into account and to consciously develop practices, policies, and regulations to reduce their emissions of harmful GHGs from, for example, landfills, transportation systems, and buildings. On the basis of the impact or effect that voluntary standards and practices have on cities and their authorities, it can be argued that they constitute normative products.

In this book, I adopt a pluralistic conception of what constitutes law and therefore use the term *law* in a broader sense.[37] It includes statements and guidelines that are not, strictly speaking, part of law but would be considered part of a broader normative or legal process. The divide between law and non-law has been the subject of long-standing discourse amongst legal theorists, and I do not intend to delve into that debate. This book situates itself firmly within the tradition that eschews a binary conception of law (i.e. an instrument is either law or it is not) and regards legal normativity as a sliding scale of varying degrees of normativity.[38] Within this tradition are the 'law as process' school and the New Haven school of international law. Former president of the International Court of Justice Rosalyn Higgins, for example, is a proponent of law as process and has argued that '[i]nternational law is not rules' or 'accumulated past decisions', but rather a continuous process from the formation of rules to their refinement through specific application by various actors, including governments, multinational corporations, international courts, and tribunals.[39] According to the New Haven school, lawmaking is a 'process of authoritative decision by which members of a community clarify and secure their common interests'.[40] It is a broad social phenomenon deeply embedded in the practices and beliefs of a society and shaped by interactions within and amongst societies.[41]

[37] On legal pluralism, see e.g. Paul Schiff Berman, *Global Legal Pluralism: A Jurisprudence of Law beyond Borders* (Cambridge University Press 2014); Balakrishnan Rajagopal, 'The Role of Law in Counter-Hegemonic Globalization and Global Legal Pluralism: Lessons from the Narmada Valley Struggle in India' (2005) 18 *Leiden Journal of International Law* 345; Gunther Teubner, '"Global Bukowina": Legal Pluralism in the World Society' in Gunther Teubner (ed), *Global Law without a State* (Dartmouth Publishing Company 1997). For a critique of the lack of continuity between 'global legal pluralism' and the older anthropological and socio-legal accounts of legal pluralism, see William Twining, 'Normative and Legal Pluralism: A Global Perspective' (2010) *Duke Journal of International and Comparative Law* 473–517.

[38] For a binary conception of law, see Jan Klabbers, 'The Redundancy of Soft Law' (1996) 65 *Nordic Journal of International Law* 167.

[39] Rosalyn Higgins, *Problems and Process: International Law and How We Use It* (Clarendon Press 1995), pgs. 2–3.

[40] Harold D. Lasswell and Myres S. McDougal, *Jurisprudence for a Free Society: Studies in Law, Science and Policy* (Martinus Nijhoff Publishers 1992), pg. xxi.

[41] Levit offers an account of 'bottom-up international lawmaking' in which lawmaking 'is a process whereby practices and behaviors gel as law' and both public and private actors 'join with others similarly situated in avocation (although often quite distant in location) to share experiences and standardize practices towards shared goals'; Janet Koven Levit, 'Bottom-Up International Lawmaking: Reflections on the New Haven School of International Law' (2007) 32 *Yale Journal of International Law* 393, pg. 409.

Adopting these conceptions of international lawmaking, this book argues that when cities construct and implement norms, practices, and voluntary standards, they are making and implementing law. The emergence of cities as jurisgenerative actors in the context of transnational climate change governance is the focus of this book.[42]

The participation of cities in transnational legal processes invites us to re-examine theories of international lawmaking that posit the state as the only legally relevant actor in international affairs. From a classical international law perspective, a city does not have international legal personality because it is deemed to be a part of the state in which it is physically and jurisdictionally embedded. Therefore, according to classical international law, the actions of cities are attributable solely to their states. If Rotterdam undertakes to reduce its GHG emissions, it simply counts towards the Netherlands' international legal obligations to mitigate climate change and does not have independent relevance for the purposes of public international law. The norms, practices, and voluntary standards that cities develop and convey through their transnational networks are also not recognized to be international law, as they are not amongst the traditional sources identified in Article 38(1) of the Statute of the International Court of Justice.

However, in the first decade of the twenty-first century, international law scholars (and those studying law and globalization more generally) increasingly recognize that we inhabit a world of multiple normative communities. After globalization, privatization, and trade liberalization swept through the world in the 1980s and 1990s, regulation and standard setting ceased to be the exclusive domain of states and international organizations. Business actors, professional associations, and NGOs have become involved in developing and implementing regulatory initiatives and voluntary codes of conduct, for example.[43] Of course, these norms have varying degrees of impact, 'but it has become clear that ignoring such normative assertions as somehow not "law" is not a useful strategy'.[44] Accordingly, what we see emerging are approaches to international law drawn from legal pluralism and transnationalism. Through its examination of cities as an emerging normative community in the sphere of transnational climate change governance, this book seeks to contribute to the larger discussions about the

[42] Steven Wheatley argues, '[t]he defining characteristic of a non-state "jurisgenerative" actor is its capacity to establish international governance norms that frame the context for action by states, corporate entities and individuals'. Further, it can be said that non-state actors exercise political authority, an activity traditionally associated with the state, when their jurisgenerative efforts have practical effect; Steven Wheatley, 'Democratic Governance beyond the State: The Legitimacy of Non-State Actors as Standard Setters' in Anne Peters et al. (eds), *Non-State Actors as Standard Setters* (Cambridge University Press 2009), pg. 220.

[43] See Section 6.2 for discussion.

[44] Paul Schiff Berman, 'A Pluralist Approach to International Law' (2007) 32 *Yale Journal of International Law* 301, pg. 302.

evolving nature of the state and international lawmaking processes in a world of increasing global pluralist governance.[45]

This book is also likely to be of interest to scholars of climate change policy. As mentioned earlier, there already exists an abundant literature on cities and climate change. There are numerous studies on the opportunities and barriers that cities face in mitigating and adapting to climate change, comparative case studies, and analyses of urban participation in hybrid governance initiatives.[46] However, to the best of my knowledge, none has considered the role of cities as norm-setters and the legal meaning of the climate governance activities in which cities engage. Neither have any studies, as far as I know, considered the interaction between the norms, practices, and voluntary standards that city networks have generated (and are putting into effect) and traditional sources of climate law such as the COP decisions of the UNFCCC and the Paris Agreement. This book therefore attempts to shed light on the transnational lawmaking dimension of the role that cities play in governing climate change. I hope that this book will enrich our understanding of a complex world prolific with climate change governance experiments that involve many public and private actors.

1.2.1 *The Focus on Global Cities*

Thousands of cities are members of networks like C40, the Covenant of Mayors for Climate and Energy,[47] the Carbon Neutral Cities Alliance,[48] and

[45] For discussion of pluralist global governance, see Grainne de Burca, Robert O. Keohane, and Charles Sabel, 'New Modes of Pluralist Global Governance' (2013) 45 *New York University Journal of International Law and Politics* 723.

[46] In addition to the works listed in 36, see e.g. Melissa Powers, 'US Municipal Climate Plans: What Role Will Cities Play in Climate Change Mitigation?' and Elizabeth Schwartz, 'Local Solutions to a Global Problem? Climate Change Policy-Making in Vancouver' in Benjamin Richardson (ed), *Local Climate Change Law: Environmental Regulation in Cities and Other Localities* (Edward Elgar Publishing 2012); Mikael Granberg and Ingemar Elander, 'Local Governance and Climate Change: Reflections on the Swedish Experience' (2007) 12 *Local Environment* 537; Benjamin J. Deangelo and L. D. Danny Harvey, 'The Jurisdictional Framework for Municipal Action to Reduce Greenhouse Gas Emissions: Case Studies from Canada, the USA and Germany' (1998) 3 *Local Environment* 111.

[47] The European Commission launched the Covenant of Mayors for Climate and Energy after the adoption of the EU Climate and Energy Package in 2008 to support local authorities in the implementation of climate mitigation and sustainable energy plans. Members of the Covenant of Mayors are eligible for funding opportunities availed by the Committee of Regions and the European Investment Bank. To become a member of the Covenant of Mayors, a local authority must give a formal undertaking to prepare a baseline emissions inventory, submit a sustainable energy action plan, and submit an implementation status report at least once every second year after submission of the sustainable energy action plan. Signatories face the possibility of suspension if they fail to submit the requisite documents within established deadlines. See the Covenant of Mayors for Climate and Energy, 'Commitment Document', online: www.covenantofmayors.eu/IMG/pdf/CoM_CommitmentDocument_en.pdf (accessed on 1 July 2016).

[48] The Carbon Neutral Cities Alliance is a project of the Urban Sustainability Directors Network, which is a 'peer-to-peer network of local government professionals' from cities across North America. On 27 March 2015, the mayors of seventeen major cities (nearly half of which are US cities, including

Eurocities,[49] to name a few. These networks have emerged to facilitate the exchange of ideas, information, and best practices amongst cities. Some networks also seek to give collective representation to urban interests and engage in political advocacy at the international level (e.g. ICLEI) and at the regional level (e.g. Eurocities). Many mid-sized cities are members of multiple networks and at some point may decide to consolidate their resources and focus on participating in networks that confer the most benefits.[50] Many cities are not likely to have the resources to participate in transnational networks that are geared towards scaling up city climate actions to the global level. For example, a vice-mayor of a mid-sized city in Greece shared in an interview that her city participated in many networks. In 2016, the mayor's office decided that it was a priority for the city to become a member of Eurocities, the regional network that is the active lobbyist for urban interests at the EU level.[51] The annual membership fee of 15,000 euros is considered hefty for a city of its size. The city's government therefore decided to withdraw from all networks that required membership fees in order to pay for the Eurocities membership. In light of this type of situation, a working assumption I adopted at an early stage of research for this book is that not all cities have the motivation and/or the resources to become globally relevant actors in climate governance.

The cities that are proactive leaders in the climate change arena tend to be those that command significant political and economic resources. These are often hubs of global trade, transnational capital, and cultural flows. In other words, the cities that are leading the current wave of urban climate action are what Saskia Sassen calls *global cities*. Sassen defines today's global cities as '(1) command points in the organization of the world economy, (2) key locations and marketplaces for the

Boston, Washington DC, and New York City) announced the launch of this alliance, which will work with member cities to achieve the goal of reducing their GHG emissions by at least 80 per cent by 2050 or sooner. See Carbon Neutral Cities Alliance, online: http://usdn.org/public/page/13/CNCA (accessed on 1 July 2016).

49 Eurocities works across a range of areas, including environmental sustainability, social cohesion, and culture. Its 'objective is to reinforce the important role that local governments should play in a multilevel governance structure' and to represent the interests of municipal authorities at the EU level. The Eurocities project team monitors relevant EU funding calls and provide members with funding forecasts, briefs, and opportunities to find project partners. Eurocities also works with EU institutions and national governments to ensure that EU legislation can be properly implemented at the local level. See Eurocities, online: www.eurocities.eu/eurocities/about_us (accessed on 1 July 2016).

50 Acuto and his colleagues have made a similar observation. They argue that the vast number of networks presents city governments with the dilemma of deciding where they should focus their networking efforts, bearing in mind that the more networks a city participates in, the greater the burden placed on an already-stretched municipal government for communication, reporting, and collaborative engagement; Michele Acuto, Mika Morissette, and Agis Tsouros, 'City Diplomacy: Towards More Strategic Networking? Learning with WHO Healthy Cities' (2017) 8(1) *Global Policy* 14.

51 Interview No. 7.

leading industries of the current period – finance and specialized services for firms, and (3) major sites of production, including the production of innovations'. According to this model, there is no such thing as a single global city or *the* global economy.[52] Global cities must interact with each other in order to fulfil the role of coordinating with and providing services to multinational corporations and even whole economies that are increasingly transnational. Such interactions amongst cities create 'inter-urban networks' or 'circuits that crisscross the world ... that connect specific areas, most of which are cities'.[53] For Sassen, 'global city' is not a descriptive term for a distinctive urban form: 'It is an analytic construct that allows one to detect the global as it is filtered through the specifics of a place, its institutional orders, and its socio-spatial fragmentations.'[54] In other words, given that the globalization processes create global cities that, in turn, are territorial spaces in which the processes of globalization unfold and affect local communities, the global city is a theoretical prism through which the connections between cities and globalization can be studied.[55] In examining the rise of cities as transnational governance actors in this book, I have broadly adopted Sassen's conception of the global city and chosen to focus on the role of global cities in governing climate change because these are the cities that are seeking to assert themselves as globally relevant actors.

1.2.2 *Defining Transnational Climate Change Governance*

Throughout the book, I will refer to the concept of transnational climate change governance. To grasp this concept requires us to engage with its three core components – transnationalism, climate change, and governance – which provide a rich vocabulary for exploring and analyzing the involvement of multiple actors (apart from states and international organizations) in governing climate change. I will briefly sketch out how each core component informs my understanding of transnational climate change governance.

Transnationalism: Transnationalism is well established in the field of international relations as the study of 'contacts, coalitions and interactions' that bridge, extend, or occur across state boundaries.[56] By definition, transnational phenomena involve

[52] Saskia Sassen, *Cities in a World Economy* (4th edn, Sage Publications 2012), pgs. 7, 111.

[53] Ibid., pg. 111. Different circuits will include different countries and cities. The circuit for the global coffee trade (including trading coffee as a commodity, selling it as a consumer product, and trading coffee futures, for example) involves cities and countries that are different from those in the circuit for biofuels production and sale. Sassen points out that the proliferation of circuits is not solely driven by economic forces. Non-governmental organizations working for labour rights and environmental protection also create and develop networks (Sassen, *Cities in a World Economy*, pg. 113).

[54] Ibid., pg. x.

[55] The concept of the global city will be elaborated upon in Chapter 2.

[56] In the 1970s, in an attempt to challenge the dominant state-centric view of world affairs, Joseph Nye and Robert Keohane sought to bring attention to the transnational dimensions of world affairs. Defining 'transnational relations' as 'contacts, coalitions and interactions across state boundaries

non-state actors (such as NGOs, businesses, charities, and religious orders) and sub-state actors (such as provinces and municipalities).[57] Whereas international affairs involve relations between states, transnational relations occur between states and non-state actors as they interact across state borders.[58]

What has been of growing interest to scholars of transnationalism in recent years are novel, hybrid governance arrangements that emerge when various groups – including business associations, NGOs, and sometimes government representatives – work together to draw up rules to induce more responsible business behaviour.[59] The plurality of actors involved in these multi-stakeholder partnerships has increased, and, as this book will show, cities now participate in some of these initiatives as partners alongside states and international organizations. An example is the Cities Climate Finance Leadership Alliance, which was launched at the UN secretary-general's Climate Summit in September 2014.[60] This global platform brings together municipal authorities, national governments, public and private financial institutions, UN agencies, businesses, and NGOs to create and implement recommendations on how to mobilize greater flow of financing towards low-emission, climate-resilient infrastructure.[61] Transnational hybrid initiatives have been the subject of a flourishing body of scholarship, which needs to be understood against the background of a wider debate on the rise of private regulation in social, economic, and environmental governance.[62] Nonetheless, it should be noted that states remain central to transnational processes. As Saskia Sassen's work has

that are not controlled by the central foreign policy organs of governments', Nye and Keohane edited a special issue of *International Organization* devoted to theorizing the impact of transnational relations on the interstate system. They also advocated a world politics paradigm that broadens the conception of actors to include non-state actors and sub-units of governments; Joseph S. Nye Jr. and Robert O. Keohane, 'Transnational Relations and World Politics: A Conclusion' (1971) 25 *International Organization* 721.

57 Harriet Bulkeley et al., *Transnational Climate Change Governance* (Cambridge University Press 2014), pg. 5. On religious organizations as transnational actors, see Jeff Haynes, 'Transnational Religious Actors and International Politics' (2001) 22 *Third World Quarterly* 143; on cities as transnational actors, see Michele Acuto, 'Global Cities as Actors: A Rejoinder to Calder and de Freytas' (2009) 29 *SAIS Review of International Affairs* 175. On regions engaging in paradiplomacy, see for example, Hubert Rioux Ouimet, 'From Sub-State Nationalism to Subnational Competition States: The Development and Institutionalization of Commercial Paradiplomacy in Scotland and Quebec' (2015) 25 *Regional and Federal Studies* 109.

58 Bulkeley et al., ibid.

59 See, for example, Luc W. Fransen and Ans Kolk, 'Global Rule-Setting for Business: A Critical Analysis of Multi-Stakeholder Standards' (2007) 14 *Organization* 667.

60 The Cities Climate Finance Leadership Alliance, 'The Bangkok-Johannesburg Blueprint', online: www.citiesclimatefinance.org/2015/12/the-bangkok-johannesburg-blueprint/ (accessed on 1 July 2016).

61 The Cities Climate Finance Leadership Alliance, 'The State of City Climate Finance 2015', online: www.citiesclimatefinance.org/wp-content/uploads/2015/12/CCFLA-State-of-City-Climate-Finance-2015.pdf (accessed on 1 July 2017). Low-emission climate-resilient infrastructure is defined as 'transport, energy, water and waste, and telecom projects that are consistent with a 2 degree Celsius pathway and resilient to the risks of climate change'; pg. 11.

62 Stefano Ponte and Carsten Daugbjerg, 'Biofuel Sustainability and the Formation of Transnational Hybrid Governance' (2015) 24 *Environmental Politics* 96, pg. 100. On the rise of private regulation, see

demonstrated, the state remains a central political entity and contributes to its own evolution through its active collaboration with and enabling of transnational forces.[63]

In the legal academy, the concept of the transnational has gained traction as lawyers grapple with the impact of globalization on the ways in which legal norms are generated, diffused, and applied across jurisdictions. In order to understand how transnational law works, one must understand transnational legal process, which Harold Hongju Koh defines as 'the transubstantive process . . . whereby states and other transnational private actors use the blend of domestic and international legal process to internalize international legal norms into domestic law'.[64] This definition helps capture the essence of contemporary legal ordering through which legal norms apply across borders and are conveyed through transnational processes.[65]

This book seeks to contribute to the transnational law literature by widening its analytical ambit to include sub-national actors such as cities. As I will argue further on, traditional conceptions of transnational law such as Koh's adopt the 'states versus private actors' dichotomy. As sub-national entities, cities cannot be accurately described as private actors, because in many ways their interests and modes of behaviour share greater similarity with states rather than private actors such as NGOs and businesses. Nijman has argued that in constituting their identities as global actors, cities imitate the foreign policy and international law practices of states in such ways as establishing foreign affairs offices and modelling UCLG, the global organization of local governments that aims to represent the world's cities at the international level, to resemble the UN.[66] One of the central claims of this book is that the unique status of the city as a sub-national actor ought to be recognized both in practice and in legal scholarship in order for a more nuanced treatment of the role of cities in transnational legal ordering to occur.

Climate Change: There is a body of sound scientific evidence for international action to address the causes and impacts of climate change. According to the IPCC's Fifth Assessment Report, we know with more than 95 per cent certainty that human

for example Tim Buthe and Walter Mattli, *The New Global Rulers: The Privatization of Regulation in the World Economy* (Princeton University Press 2011).

63 'The national is still the realm where formalization and institutionalization have all reached their highest level of development', but 'the national is also often one of the key enablers and enactors of the emergent global scale'; Saskia Sassen, *Territory, Authority, Rights: From Medieval to Global Assemblages* (Princeton University Press 2008), pg. 3.

64 Harold H. Koh, 'Why Transnational Law Matters' (2005–2006) 24 *Penn State International Law Review* 745, pgs. 745–746.

65 Gregory Shaffer, 'A Transnational Take on Krisch's Pluralist Postnational Law' (2012) 23 *European Journal of International Law* 565, pg. 577; also see Gregory C. Shaffer (ed), *Transnational Legal Ordering and State Change* (Cambridge University Press 2014).

66 Janne E. Nijman, 'Renaissance of the City as Global Actor: The Role of Foreign Policy and International Law Practices in the Construction of Cities as Global Actors' in Gunther Hellmann, Andreas Fahrmeir, and Milos Vec (eds), *The Transformation of Foreign Policy: Drawing and Managing Boundaries from Antiquity to the Present* (Oxford University Press 2016).

activity is responsible for the increase in global average temperature.[67] The emission of GHGs, particularly carbon dioxide, is the dominant cause of the observed warming since the mid twentieth century.[68] Economic and population growth are the most important drivers of increases in carbon dioxide emissions, mainly from the burning of fossil fuels. Due to human activity, the atmospheric concentrations of carbon dioxide, methane, and nitrous oxide have all increased since 1750.[69] Cities currently account for over 70 per cent of global energy use and are therefore a key source of GHG emissions.[70] The 'carbon footprint' of cities will continue to increase if efforts are not taken to control and reduce urban emissions as urbanization is on the rise.[71] It is expected that the number of people living in cities in 2050 will be 2.5 to 3 billion greater than in 2009.[72]

Since the 2009 UNFCCC COP in Copenhagen, the target to limit average global temperature rise to no more than 2 °C above pre-industrial levels has become firmly entrenched in international climate policy discourse.[73] In the Paris Agreement, signatory states committed to strengthen the global response to climate change by 'holding the increase in the global average temperature to well below 2 °C above pre-industrial levels and pursuing efforts to limit the temperature increase to 1.5 °C above pre-industrial levels, recognizing that this would significantly reduce the risks and impacts of climate change'.[74] Deep cuts in global GHG emissions are required to achieve this goal. In fact, halting the global average temperature rise at any level will require net zero global carbon dioxide emissions at some point in the future. The IPCC, the UN Environment Programme (UNEP), and the UNFCCC have repeatedly emphasized that the longer we wait to bend the currently increasing curve of global emissions downwards, the steeper we will have to bend it.[75] Limiting global warming to below 2 °C 'necessitates a radical transition

[67] T. F. Stocker et al., *IPCC, 2013: Summary for Policymakers in Climate Change 2013: The Physical Science Basis. Contribution of Working Group I to the Fifth Assessment Report of the Intergovernmental Panel on Climate Change* (Cambridge University Press 2013), pg. 15.

[68] Ibid., pg. 17. [69] Ibid., pg. 11.

[70] Cambridge Institute for Sustainability Leadership, Cambridge Judge Business School, and ICLEI, *Climate Change: Implications for Cities (Key Findings from the Intergovernmental Panel on Climate Change Fifth Assessment Report)* (2014), pg. 5, online: www.cisl.cam.ac.uk/ipcc (accessed on 1 July 2016).

[71] Ibid. [72] Ibid.

[73] For discussion of the history of the two-degree Celsius target, see Samuel Randalls, 'History of the 2C Climate Target' (2010) 1 *WIREs Climate Change* 598. In the first paragraph of the Copenhagen Accord, signatories pledge that '[t]o achieve the ultimate objective of the Convention to stabilize greenhouse gas concentration in the atmosphere at a level that would prevent dangerous anthropogenic interference with the climate system, we shall, recognizing the scientific view that the increase in global temperature should be below 2 degrees Celsius … enhance our long-term cooperative action to combat climate change'; Decision 2/CP.15.

[74] Article 2(1)(a) of the Paris Agreement.

[75] UNFCCC, *Report on the Structured Expert Dialogue on the 2013–2015 Review: Note by the Co-Facilitators of the Structured Expert Dialogue* (FCCC/SB/2015/INF1, 4 May 2015), pg. 11.

(deep decarbonization now and going forward), not merely a fine tuning of current trends'.[76]

In this book, climate change is broadly framed as an environmental problem caused by socio-economic disruptions. The solutions require fundamental economic and social transformation such that climate change governance must take place at multiple scales of governance and involve many actors. This includes cities and their governments implementing solutions at the sub-national level and scaling up their actions through transnational cooperation. In general, climate action can be divided into two main categories: mitigation and adaptation. *Mitigation* refers to measures to reduce GHG sources and remove GHGs from the atmosphere.[77] Examples include improving the energy efficiency of buildings so that they use less electricity and planting trees to absorb atmospheric carbon. The IPCC defines *adaptation* as '[a]djustment in natural or human systems in response to actual or expected climatic stimuli or their effects, which moderates harm or exploits beneficial opportunities'.[78] Examples include strengthening a city's defences against flooding risks and improving the quality of road surfaces to withstand higher temperatures. Mitigation and adaptation measures can often lead to synergies.[79] For example, increasing tree cover in a city not only mitigates climate change;[80] it also cools the city, reducing the likelihood of heatstroke as city inhabitants face hotter summers. Increasing a city's tree cover also delivers a range of other benefits, including improved air quality and more biological diversity.[81]

Governance: At the most general level, governance may be understood as 'issues of social coordination and the nature of all patterns of rule'.[82] This level of generality and definitional flexibility is one of the strengths of the governance concept, as it allows for the exploration of phenomena that do not fit well into established categories. However, such a broad definition also occasionally makes it difficult to define what, exactly, constitutes governance. Further, there are multiple definitions of governance in the literature. For example, Michael Zurn et al. define governance as the sum of regulations, policy processes, and structures which are designed to remedy a public problem via a collective course of action.[83] Pattberg and Stripple would consider governance as a matter of shaping the conduct of actors through the

76 Ibid.

77 IPCC, 'Climate Change 2007: Working Group II: Impacts, Adaptation and Vulnerability', Glossary, online: www.ipcc.ch/publications_and_data/ar4/wg2/en/annexessglossary-a-d.html

78 Ibid.

79 See, for example, AMICA, online: www.amica-climate.net/about_amica.html (accessed on 1 July 2016).

80 Kenton Rogers et al., *Valuing London's Urban Forest: Results of London i-Tree Eco Project* (Treeconomics London 2015).

81 Ibid., pgs. 45–50.

82 Mark Bevir, *The SAGE Handbook of Governance* (SAGE Publishing 2011), pg. 1.

83 Michael Zurn et al., 'Introduction' in Henrik Enderlein, Sonja Walti, and Michael Zurn (eds), *Handbook on Multi-Level Governance* (Edward Elgar Publishing 2010), pg. 2.

establishment of new norms and practices such as carbon reporting.[84] For the purposes of this book, I adopt Andonova et al.'s definition of (transnational) governance: 'when networks operating in the transnational sphere authoritatively steer constituents toward public goals'.[85] There are three elements to this definition that are worth noting. First, it is concerned with realizing public goals. Second, steering connotes the directive and intentional nature of governance processes. Third, governance is regarded as authoritative, defined as 'the ability to induce deference in others'.[86]

In summary, transnational climate change governance occurs when sub-state actors, private actors, states, and international organizations engage in cross-border cooperation, thereby forming transnational networks to transmit knowledge, best practices, and expertise. Transnational climate change governance seeks to authoritatively steer and have a constraining impact on the behaviour of target actors. When norms, practices, and voluntary standards emerge, it can be argued that transnational climate change governance takes on a normative dimension. Based on a pluralistic conception of lawmaking, the production of norms and voluntary standards that are usually adopted and implemented through practices can be viewed as law. Transnational climate change governance is specifically concerned with constraining the GHG-emitting behaviour of target actors, with the goal of limiting average global temperature rise to no more than 2 °C above pre-industrial levels. It is also concerned with building the resilience of communities to the impacts of climate change such as increased flood risks.

1.2.3 *Research Questions*

Summing up the discussion so far, the central claim of this book is that cities are beginning to perform lawmaking functions in the context of transnational climate change governance. Through transnational networks that form important linkages between city governments, states, international organizations, businesses, and civil society, cities are creating and implementing norms, practices, and voluntary standards across national boundaries. The impact of their normative output is to steer the behaviour of cities towards mitigating climate change and adapting to its unavoidable impacts. Further, the multi-level, multi-actor partnerships in which cities participate actively today transcend the public-private divide and the traditional strictures of 'domestic' (internal) versus 'international' (external). Hence, the

[84] Philipp Pattberg and Johannes Stripple, 'Beyond the Public and Private Divide: Remapping Transnational Climate Governance in the 21st Century' (2008) 8 *International Environmental Agreements* 367.

[85] Liliana B. Andonova, Michele M. Betsill, and Harriet Bulkeley, 'Transnational Climate Governance' (2009) 9 *Global Environmental Politics* 52, pg. 56.

[86] Avant, Finnemore, and Sell, pg. 9. Deference to authority can take various forms. In the present case, the authority of 'climate leaders' has tended to create new preferences or persuade other city leaders/ mayors to share the leaders' views on climate change and the possible solutions.

transnational governance activities of cities also point to the reconfiguration of the state in modern times – from a monolithic, unitary entity to a mutable, variegated one in a global system densely populated by networks of components of the state (for example, networks comprising judges, antitrust enforcement agencies, and central banks).[87]

To examine and establish the above claims, the following questions guide this research agenda:

1. What recent developments suggest the rise of cities in international affairs?
2. What have cities been doing to govern climate change, and which of these governance activities generate normative effects transnationally?
3. How do the norms, practices, and voluntary standards generated by cities and transmitted by their networks lead to cities reducing their GHG emissions and increasing their climate resilience?
4. How do the norms, practices, and voluntary standards generated by cities and transmitted by their networks relate to those of the UNFCCC regime?
5. Do global cities make a positive normative contribution to the global constellation of climate change governance activities?

These questions are complemented by sub-questions in each chapter.

1.3 CHAPTER OUTLINE

This book consists of eight chapters, including this introduction chapter and the conclusion. Chapter 2 sets out an eclectic analytical framework that draws upon diverse theoretical perspectives from various disciplines – including political science, law, international relations, and sociology – to help frame the participation of cities in transnational climate change governance processes. Transnational regime complex theory provides the framing of transnational climate change governance as a space that is densely populated with many institutions that perform potentially overlapping functions. Cities are emerging within this space as one of multiple actors. After introducing regime complex theory, the chapter moves on to briefly explain the theory of orchestration, which is primarily prescriptive about how the international system ought to manage the disadvantages and maximize the benefits of institutional fragmentation.

Transnational law provides an alternative theoretical framework for analyzing how states and non-state actors are involved in making and implementing law that has effect transnationally. The shift from a state-centric view of international law is necessary in order to consider the role of sub-state actors such as cities in governing climate change at the global level. Anne Marie Slaughter's theory of the

[87] See Anne-Marie Slaughter, *A New World Order* (Princeton University Press 2005). This point will be further explored in Chapter 7.

disaggregated state serves as a point of departure in this book for conceptualizing cities as entities that are capable of participating in transnational governance. This book also borrows Slaughter's idea of transgovernmental networks and adapts it to conceptualize city networks. Finally, the body of literature that theorizes the global city as an actor in world affairs is an important cornerstone of the theoretical framework for this book because it is upon this idea that I develop my argument that cities are emerging as lawmaking actors.

Chapter 3 is titled 'The Rise of the City in International Affairs'. This chapter situates the emergence of cities as participants in transnational climate change governance within the broader context of how cities increasingly participate directly and independently in international affairs and global politics. It starts out by describing instances of cities developing what appear to be their own independent foreign and security policies despite the fact that such policy areas have traditionally been considered the core domains of central governments and states. It then considers instances of cities implementing international treaties (for example, the Paris Agreement) on their own accord, sometimes in opposition to the stance adopted by their states, thereby challenging traditional conceptions of how international law is internalized and implemented domestically. Cities have also formed organizations to represent their interests in international forums and to achieve foreign policy objectives such as nuclear non-proliferation.

The final part of this chapter examines how the 'internationalization of cities' has gained traction because of a shift in the development agenda of international organizations, including the World Bank and UN-Habitat, towards sustainable urban development. The contemporary global urban agenda has created programmes and mechanisms that link cities directly to international organizations, bypassing national governments. These programmes also channel resources directly to cities, which enhances their position vis-à-vis their central government. While the practical effects of these urban programmes and policies are mostly experienced in cities in developing countries and least developed countries, as these cities are usually the recipients of aid and technical assistance from international organizations, this book argues that the global urban agenda has strong ideological influence and creates a normative environment that is conducive for cities to exercise agency in global governance.

Chapter 4 examines what five global cities – London, Mexico City, New York City, Rotterdam, and Seoul – are doing to address climate change. These cities are internationally recognized by policymakers, city planners, the media, international organizations, and researchers to be 'climate frontrunners'. The aim of this chapter is to provide an overview that will give readers a sense of what global cities are doing 'on the ground' within their territories to govern climate change.

An appreciation of the localized practices and practical aspects of global city action on climate change is important in at least two respects. One of the key ideas this book seeks to advance is that the practices, policies, and strategies taken at the city level contribute towards transnational climate change governance when global cities seek to scale up their actions and pursue cooperation through cross-border networks. In this chapter, one observes the intertwined connections between municipal institutions and global organizations, between local practices and transnational norms. Furthermore, this chapter highlights the fact that a number of commonalities unite these climate frontrunner global cities. The features that these cities share include (1) visionary leadership by a mayor who enjoys the support of the city's administrative agencies, the private sector, civil society, and other major stakeholders; (2) public support for ambitious climate action; and (3) active participation in global networks such as C40. This helps put into perspective the role that global cities can play in the transnational climate change regime complex. While all cities, large and small, may want to contribute to the global effort to mitigate climate change, only a select few global cities with the resources, political will, and cosmopolitan vision of the common good will be active participants in transnational governance and rule-making processes.

Chapter 5 posits that cities do not act alone. They form networks to aggregate their bargaining power vis-à-vis sellers of low-carbon technologies, facilitate cooperation, develop voluntary standards, and convey norms. These networks often establish important links with other actors at other scales [e.g. regional organizations such as the Association of Southeast Asian Nations (ASEAN)]. The network structure and its links with actors at multiple governance levels provide channels for cities to diffuse norms, practices, and voluntary standards.

C40 is the most prominent network of global cities committed to addressing climate change and is the subject of detailed analysis in this chapter. Briefly, C40 brands itself as a gathering of the top echelon of the world's major cities. A key aspect of C40's governance mode is to facilitate corporation between its member cities and other transnational actors. This includes establishing and coordinating public-private partnerships with multinational corporations like Siemens, NGOs such as the World Resource Institute, and international financial institutions like the World Bank.

Drawing on the descriptive and analytical examination in the previous three chapters, Chapter 6 advances the argument that cities are emerging as transnational lawmakers. The chapter begins by identifying the norms that global cities have developed and internalized through reiterative interaction and frequent cooperation within their network. These norms are as follows: first, that climate change is a global problem but *can* and *must* be addressed locally by cities; second, that large, global cities are not only a source of the problem because of their high levels of GHG emissions, but also a source of solutions; third, that cities can best reduce their GHG emissions and embark on low-carbon growth by committing to a policy approach based on measurement, monitoring, and reporting of their GHG emissions. Based on this normative foundation, global cities have cooperated (through C40) with other actors such as consultancies, development banks, and civil society to develop practices and voluntary standards to enable global cities to reduce their GHG emissions in the short term and make the transition towards low-carbon development in the longer term. I coin the term *urban climate law* to refer to the norms, practices, and voluntary standards created by global cities and implemented through their transnational networks.

Chapter 6 also identifies the promotion of reflexivity and norm diffusion as the two main pathways of influence that are critical for steering cities towards climate mitigation and low-carbon development. Finally, this chapter advances the claim that, in many key respects, urban climate law has been deliberately designed to support and reinforce the UNFCCC regime. The term 'coupling' is used to refer to the deliberate effort to align urban climate law in ways such that its norms and practices complement and strengthen the UNFCCC regime. I argue that coupling not only benefits the UNFCCC regime but also supports the development of urban climate law by conferring legitimacy by association and implicit endorsement by the UNFCCC regime. The analysis on coupling also responds to broader criticisms against soft law by demonstrating that the soft law generated by cities is an important complement to hard law that can facilitate experimentation at multiple sites and levels of governance, generate knowledge, build trust, and transform norms.

Chapter 7 endeavours to take a step back and consider some 'big picture' questions. The first question explored in this chapter is how the transnational cooperative efforts among global cities and urban climate law contribute towards the performance of the transnational climate change regime complex. After concluding that global cities and their networks have a valuable normative role to play in the climate change regime complex, I take a step further in the macro-level analysis and consider what the emergence of cities as transnational lawmaking actors means for the study of international law and international relations more broadly. An argument advanced in this chapter is that the study of cities and their normative activities reinforces the idea that international law no longer refers solely or even primarily to the law that governs the rights and obligations of states in co-existence and that there are various theoretical conceptions of the international/transnational/global legal order. International legal scholarship is enriched by the recognition of these various schools of thought and a move away from state-centric versions of international lawmaking, which hinder the crafting of creative solutions to global collective action problems such as climate change. This chapter also considers how the responses of international law and practice to the emergence of global cities contribute towards the ongoing multidisciplinary conversation about global cities.

Chapter 8 concludes the book with a series of reflections.

2

Theoretical Framework

2.1 INTRODUCTION

In order to generate a richer understanding of the emergence of cities as governors and, more specifically, lawmaking actors in the area of transnational climate change governance, this book adopts an eclectic analytical framework that draws upon diverse theoretical perspectives. Analytical eclecticism proceeds on the basis of a pragmatic ethos. It is said that the researcher who adopts an analytical eclectic approach is searching for middle-range theoretical arguments that potentially speak to concrete issues of policy and practice.[1] In drawing upon diverse theoretical perspectives, this book seeks to guard against excessive simplification that can arise when one tries to apply a single theoretical lens to explain messy real-world situations. In other words, analytical eclecticism invites us to refrain from being 'intellectually aggressive hedgehogs'[2] and to embrace the possibilities that open up from drawing upon theories and narratives developed in different research traditions.

This chapter provides the theoretical underpinnings for the book by bringing together theories from various disciplines including political science, law, international relations, and sociology to help frame the participation of cities in transnational climate change governance processes. Transnational regime complex theory provides the framing of transnational climate change governance as a space that is densely populated with many institutions that perform potentially overlapping functions. Cities are emerging within this space as one of multiple governance actors. After introducing the regime complex theory, the chapter moves on to briefly explain the theory of orchestration, which is primarily prescriptive about how the international system ought to manage the disadvantages and maximize the benefits of institutional fragmentation. While regime complex theory tends to emphasize the problems of rule inconsistency and rule conflict that can arise when there are multiple institutions governing an issue area, orchestration theory emphasizes the

[1] Rudra Sil and Peter J. Katzenstein, 'Analytic Eclecticism in the Study of World Politics: Reconfiguring Problems and Mechanisms across Research Traditions' (2010) 8 *Perspectives on Politics* 411.

[2] Ibid., pg. 414.

benefits of institutional multiplicity such as the potential for mutual learning and experimenting with different governance approaches.

Transnational law provides an alternative theoretical framework for analyzing how states and non-state actors are involved in making and implementing law that has effect across national boundaries. Making a shift from a state-centric view of international law is necessary in order to consider the role of sub-state actors such as cities and their local governments in governing climate change at the global level. Anne Marie Slaughter's theory of the disaggregated state and transgovernmental networks serves as a point of departure in this book for conceptualizing the possibility of cities as entities that are capable of participating in transnational governance processes. This book also borrows the idea of transgovernmental networks and adapts it to conceptualize city networks as a tool of governance in the transnational climate change regime complex. Finally, the body of literature that theorizes the global city as an actor in world affairs is an important cornerstone of the theoretical framework for this book because it underpins the argument that cities are emerging as law-making actors.

2.2 THEORETICAL OVERVIEW

2.2.1 *Transnational Regime Complex*

From the mid-1990s onwards, patterns of institutionalization in global governance changed. Efforts to develop new comprehensive and integrated international regimes failed, while established ones began to fragment. For example, intensive negotiations amongst OECD members over a three-year period on what was to be a Multilateral Agreement on Investment (MAI) eventually collapsed.[3] Apart from substantial disagreement amongst the negotiating states, a global coalition of environmental and development NGOs, citizen groups, and governments of developing countries successfully put pressure on OECD governments to withdraw from the negotiations.[4]

After the creation of the World Trade Organization (WTO), officials had hoped to start what is now known as the Doha Round of negotiations during the Seattle WTO

[3] The documents relating to the MAI negotiations between 1995 and 1998 can be found in an online database: OECD, 'Multilateral Agreement on Investment: Documents from the Negotiations', www1.oecd.org/daf/mai/intro.htm (accessed on 1 July 2016).

[4] Those who opposed the MAI argued that, amongst other things, the agreement would lead to a 'race to the bottom' in environmental and labour standards. See, for example, Andrea Durbin and Mark Vallianatos, *Transnational Corporate Bill of Rights – Negotiations for a Multilateral Agreement on Investment (MAI)* (Friends of the Earth 1997), online: www.globalpolicy.org/component/content/article/209/43203.html; Martin Khor, *NGOs Mount Protests against MAI* (Third World Network Features 1998), online: www.globalpolicy.org/component/content/article/209-bwi-wto/43217.html (accessed on 1 July 2016). For discussion, see, for example, Eric Neumayer, 'Multilateral Agreement on Investment: Lessons for the WTO from the Failed OECD-Negotiations' (1999) 46 *Wirtschaftspolitische Blätter* 618.

ministerial talks in November 1999.[5] The objective of these negotiations was to lower trade barriers and therefore facilitate global trade. However, this did not happen because the Seattle meeting collapsed in the face of anti-globalization protests that escalated into riots, looting, and the use of tear gas to quell the protesters.[6] Negotiations on the package of agreements – known as the Doha Development Agenda – eventually commenced in 2001, but WTO Director-General Pascal Lamy suspended negotiations in July 2006 because of irreconcilable differences between WTO members.[7] Since then, efforts have been under way to resume negotiations; however, there is little optimism about achieving a breakthrough.[8] The impasse in the Doha Round has called the multilateral approach into question, with some arguing that WTO membership has grown so large and the issues that have been taken on are so complex that it is almost impossible to reach consensus.[9]

A final example would be the international climate change negotiations. At the first UNFCCC COP held in Berlin in 1995, member states agreed that the absence of legally binding GHG emission reduction targets rendered the UNFCCC a relatively weak legal instrument.[10] They therefore agreed to negotiate a protocol that would include binding targets.[11] This led to the Kyoto Protocol, which was adopted on 11 December 1997.[12] In line with the principle of common but differentiated responsibilities and respective capabilities (CBDRRC), the Kyoto Protocol required developed countries to adopt binding targets, but developing countries were not required to do so. Without emission targets imposed on China and other rapidly developing countries, the US Congress strongly opposed the Kyoto Protocol and refused to ratify it.[13] Without US ratification, the protocol could only enter into

[5] World Trade Organization, 'The Doha Agenda', online: www.wto.org/english/thewto_e/whatis_e/tif_e/doha1_e.htm (accessed on 1 July 2016).

[6] John Vidal, 'Real Battle for Seattle' *Guardian* (5 December 1999). Influential environment and human rights groups from both the Global North and Global South condemned how the talks were being conducted. They alleged that genuine concerns about the effects of another round of trade liberalization on the environment, jobs, and cultural issues were subordinated to pure economic interests and that governments of Third World countries were marginalized during the talks.

[7] World Trade Organization, 'The Doha Agenda'.

[8] See, for example, Jean-Pierre Lehmann, 'End the Charade in Talks on Global Trade' *Financial Times* (24 August 2011). Lehmann suggests that 'the Doha round should be buried. Some suggest it should be declared dead. But it has been dead for some time and the corpse is putrefying: so a burial, a wake, and some appropriate words of farewell.' For a less pessimistic view, see Jagdish Bhagwati, 'From Seattle to Hong Kong' (2005) 12 *Foreign Affairs*, online: www.foreignaffairs.com/articles/2005-12-01/seattle-hong-kong (accessed on 3 July 2016).

[9] See, for example, Razeen Sally, 'The End of the Road for the WTO? A Snapshot of International Trade Policy after Cancun' (2004) 5 *World Economics* 1; Alan Beattie, 'The Multilateral Approach Is Called into Question' *Financial Times* (15 November 2005).

[10] UNFCCC Decision 1/CP. 1, preamble. [11] Ibid.

[12] 1997 Kyoto Protocol to the United Nations Framework Convention on Climate Change, 2303 UNTS 148. The individual targets for Annex I parties are listed in Annex B of the Kyoto Protocol.

[13] The Byrd-Hagel Resolution (US Senate Resolution 98), 105th Congress, 1st session, 25 July 1997, states that 'the exemption for Developing Country Parties is inconsistent with the need for global action on climate change and is environmentally flawed' and that the differentiated treatment of Annex I parties

force in 2005.[14] Further, without the participation of the United States, one of the world's largest GHG emitters as well as a global superpower, the Kyoto Protocol regime was hampered from the outset. The negotiations for the post-Kyoto framework, which finally culminated in the Paris Agreement, have been similarly fraught with difficulties as developed and developing countries struggled to find accommodation over crucial issues such as financial assistance to the developing countries to cope with climate change.[15]

Against this background of dispersed power, disparate interests amongst states, and the inertia of established international organizations, new alternative governance arrangements started to emerge from the 1990s onwards. One of these novel governance arrangements has been termed *regime complex*. As originally defined in the international relations scholarship, a regime is a set of implicit or explicit principles, norms, rules, and decision-making procedures around which actors' expectations converge in an issue area such as climate change.[16] A regime complex emerges when

and non-Annex parties 'could result in serious harm to the United States economy, including significant job loss, trade disadvantages, increased energy and consumer costs, or any combination thereof'. The Senate passed the resolution unanimously (95–0); Council on Foreign Relations, Byrd-Hagel Resolution, online: www.cfr.org/climate-change/byrd-hagel-resolution/p21331 (accessed on 12 August 2015). For discussion, see Daniel A. Farber, 'Climate Justice and the China Fallacy' (2009) 15 *Hastings West-Northwest Journal of Environmental Law and Policy* 15.

[14] Pursuant to Article 25, the Kyoto Protocol could only enter into force upon ratification by 'not less than 55 Parties to the Convention, incorporating Parties included in Annex I which accounted in total for at least 55 per cent of the total carbon dioxide emissions for 1990 of the Parties included in Annex I'. Without the United States – the world's largest GHG emitter at that time – on board, it was only after Russia ratified the agreement that the Kyoto Protocol finally entered into force on 16 February 2005. Russia's ratification was allegedly the result of a bargain struck with the EU concerning Russia's accession to the World Trade Organization. Andrey Illarionov, President Vladimir Putin's then economic policy adviser, went so far as to characterize the Kyoto Protocol as an 'economic Auschwitz' for Russia and 'an assault on science, economic growth and human freedom'. However, despite internal disagreements, once the decision was made at the political level, Russia deposited its instrument of ratification with the UN within weeks; Yulia Yamineva, 'Climate Law and Policy in Russia: A Peasant Needs Thunder to Cross Himself and Wonder' in Erkki J. Hollo, Kati Kulovesi, and Michael Mehling (eds), *Climate Change and the Law* (Springer 2013), pgs. 553–554.

[15] For a sample of the literature on the negotiations for the post–Kyoto Protocol framework, see Meinhard Doelle, 'The Legacy of the Climate Talks in Copenhagen: Hopenhagen or Brokenhagen?' (2010) 1 *Carbon and Climate Law Review* 86; Matthieu Wemaere, 'State of Play of International Climate Negotiations: On the Road to Copenhagen' (2009) 4 *Carbon and Climate Law Review* 497; Richard Black, 'Why Did Copenhagen Fail to Deliver a Climate Deal?' *BBC* (22 December 2009); Francesco Sindico, 'The Copenhagen Accord and the Future of the International Climate Change Regime' (2010) 1 *Revista Catalana de Dret Ambiental* 1; Harro van Asselt, Francesco Sindico, and Michael Mehling, 'Global Climate Change and the Fragmentation of International Law' (2008) 30 *Law & Policy* 423; Lavanya Rajamani, 'The Cancun Climate Agreements: Reading the Text, Subtext and Tea Leaves' (2011) 60 *International and Comparative Law Quarterly* 499; Remi Moncel, 'Unconstructive Ambiguity in the Durban Climate Deal of COP 17/CMP 7' (2012) 12 *Sustainable Development Law & Policy* 6; Pilita Clark, 'Climate Talks Struggle into Overtime' *Financial Times* (13 December 2014);Pilita Clark, 'UN Climate Agreement Reached in Marathon Session' *Financial Times* (14 December 2014).

[16] Stephen D. Krasner, 'Structural Causes and Regime Consequences: Regimes as Intervening Variables' (1982) 36 *International Organization* 185, pg. 2.

an issue area is no longer governed by a single regime. Instead, the relevant rules are found in a number of regimes that overlap in their scope and subject matter. In their pioneering article, Raustiala and Victor define a regime complex to be 'an array of partially overlapping and non-hierarchical institutions governing a particular issue-area'.[17] The issue of overlap has been a core concern of regime complex theory. The existence of multiple institutions can lead to regime shifting, whereby actors move issues to forums that most suit their strategic interests.[18] Furthermore, when two or more regimes create rules that govern the same issue, there is the risk of inconsistent rules. Rule inconsistency creates uncertainty and costs for institutions and their members, particularly for intergovernmental organizations that administer legally binding rules. International legal scholarship on proliferation of institutions tends to focus on the question of coherence or, viewed from a different angle, on 'fragmentation', which is understood as potential for 'conflicts between rules or rule-systems, deviating institutional practices and, possibly, the loss of an overall perspective on the law'.[19]

Keohane and Victor have offered a broader definition of a regime complex in the case of climate change, whereby a regime complex comprises of a collection of loosely linked regimes that are '*sometimes conflicting, usually mutually reinforcing*' (my emphasis).[20] This definition departs from the traditional one, which emphasizes overlap in the regimes. Keohane and Victor argue that, in the case of climate change, the diverse interests amongst states make it more likely that a regime complex emerges instead of a single, integrated, and comprehensive regime for

[17] Kal Raustiala and David G. Victor, 'The Regime Complex for Plant Genetic Resources' (2004) 58 *International Organization* 277, pg. 279. Orsini et al. have proposed an alternative definition of a regime complex. In their view, Raustiala and Victor's definition has 'several ambiguous features that impede further analysis'. They therefore define a regime complex as 'a network of three or more international regimes that relate to a common subject matter; exhibit over-lapping membership; and generate substantive, normative or operative interactions recognized as potentially problematic whether or not they are managed effectively'; Amandine Orsini, Jean Frédéric Morin, and Oran Young, 'Regime Complexes: A Buzz, a Boom, or a Boost for Global Governance?' (2013) 19 *Global Governance* 27, pg. 29.

[18] For discussion, see, for example, Laurence R. Helfer, 'Regime Shifting in the International Intellectual Property System' (2009) 7 *Perspectives on Politics* 39; Laurence R. Helfer, 'Regime Shifting: The TRIPS Agreement and New Dynamics of International Intellectual Property Lawmaking' (2004) 29 *Yale Journal of International Law* 1.

[19] Martti Koskenniemi, *Fragmentation of International Law: Difficulties Arising from the Diversification and Expansion of International Law: Report of the Study Group of the International Law Commission* (UN General Assembly, A/CN4/L682, 2006), para. 8. Simma argues that fragmentation is not a crisis but a sign of the growing maturity of the international legal order; Bruno Simma, 'Fragmentation in a Positive Light' (2004) 25 *Michigan Journal of International Law* 847. Dupuy and Viñuales are of the view that the challenge is not one of fragmentation but one of 'integration' whereby differences in legal interpretation are resolved by the development of principles capable of integrating different solutions within a common framework; Pierre-Marie Dupuy and Jorge E. Viñuales, 'The Challenge of "Proliferation": An Anatomy of the Debate' in Cesare Romano, Karen J. Alter, and Yuval Shany (eds), *The Oxford Handbook of International Adjudication* (Oxford University Press 2014), pgs. 148–149.

[20] Robert O. Keohane and David G. Victor, 'The Regime Complex for Climate Change' (2011) 9 *Perspectives on Politics* 7, pg. 7.

managing climate change.[21] Keohane and Victor's mapping of the regime complex for climate change, in line with traditional regime complex theory, focuses on states and international organizations. Subsequently, Abbott has contended that Keohane and Victor's conception of the climate change regime complex provides an incomplete picture of global climate change governance because it excludes a range of governance initiatives that involve non-state actors (such as environmental NGOs, technical experts, and business associations) and operate transnationally.[22] In advancing the notion of a transnational regime complex for climate change, Abbott refined Keohane and Victor's conception by adding the transnational dimension – i.e. the involvement of private actors and sub-state actors in developing and implementing governance initiatives that often have cross-border effects. Thus, a transnational regime complex for climate change includes international organizations and treaty bodies such as the UNFCCC secretariat, World Bank Prototype Carbon Fund, and Montreal Protocol on Substances that Deplete the Ozone Layer, *as well as* private and hybrid (public-private) governance initiatives such as the World Business Council for Sustainable Development, Asian Cities Climate Change Resilience Network, and International Emissions Trading Association.[23] In this book, I adopt Abbott's conception of the transnational climate change regime

[21] Ibid.

[22] Kenneth W. Abbott, 'The Transnational Regime Complex for Climate Change' 30 *Environment and Planning C: Government and Policy* 571, pg. 574.

[23] The UNFCCC secretariat describes its functions as follows: 'In its early years, the main task of the secretariat was to support intergovernmental climate change negotiations. . . . Currently, a major part of our work involves the analysis and review of climate change information and data reported by Parties.'; UNFCCC, 'The Secretariat', online: http://unfccc.int/secretariat/items/1629.php (accessed on 1 April 2016). The World Bank Prototype Carbon Fund is the first carbon fund in the world. Its mission is to pioneer emissions trading on a global scale. Since it became operational in April 2000, the fund has been active in promoting the Kyoto Protocol's Clean Development Mechanism and Joint Implementation; Charlotte Streck, 'World Bank Carbon Finance Business: Contracts and Emission Reductions Purchase Transactions' in David Freestone and Charlotte Streck (eds), *Legal Aspects of Implementing the Kyoto Protocol Mechanisms: Making Kyoto Work* (Oxford University Press 2005), pg. 370. The Montreal Protocol on Substances That Deplete the Ozone Layer 1522 UNTS 3 (1987) is designed to reduce the production and consumption of substances that deplete ozone in the stratosphere. Ozone-depleting substances include chlorofluorocarbons (CFCs) and hydrochlorofluorocarbons (HCFCs). CFCs and HCFCs are also greenhouse gases. Hydrofluorocarbons (HFCs) and perfluorocarbons (PFCs) are used as substitutes for CFCs and HCFCs in some applications because they do not deplete the ozone layer. However, these substitutes (i.e. HFCs and PFCs) are greenhouse gases that the Kyoto Protocol aims to reduce. Thus, the regulatory choices and decisions of the Montreal Protocol regime have implications for climate change governance, and the respective treaty secretariats have taken steps towards coordination and information sharing. See UNFCCC, 'Methodological Issues Relating to Hydrofluorocarbons and Perfluorocarbons', online: http://unfccc.int/methods/other_methodological_issues/items/2311.php (accessed on 1 August 2016). The World Business Council for Sustainable Development (WBCSD) brings together companies that seek to develop and implement solutions to address environmental and sustainability challenges. WBCSD has a number of major projects related to cities. For example, it has issued a set of Sustainable Mobility Indicators, which provides cities with a diagnostic tool to assess their performance and develop sustainable urban mobility plans; WBCSD, 'The European Commission Endorses WBCSD Set of Indicators to Help Cities Advance Sustainable Mobility', online: www

complex as a framework for describing and analyzing the involvement of cities and their networks in global climate change governance. Unlike Keohane and Victor, Abbott's definition opens up the conceptual space that is necessary for considering a larger cast of actors, apart from states and international organizations, in governing climate change. It also pays more attention to the causes and effects of having multiple regimes involved in climate change governance and how existing regimes interact, but it downplays the traditional focus on overlapping norms and rules.[24] At the same time, because Abbott's definition builds upon traditional regime complex theory, this book is able to draw on a rich body of literature on institutional multiplicity that offers relevant insights about transnational climate change governance even though regime complex theory was developed with states in mind.

For example, regime complex theory identifies the benefits and drawbacks of having multiple actors involved in governance. Regarding advantages, Keohane and Victor argue that regime complexes offer flexibility across issues and adaptability over time. On the aspect of flexibility, they suggest that when there is no requirement that all rules be bound within a single institution, it opens up the possibility of adapting rules for different coalitions of actors and fine-tuning the application of rules for different conditions.[25] As will be discussed in Chapter 6, flexibility has been an advantage that permits cities to create norms, practices, and voluntary standards that are more responsive to urban needs and interests. This increases the likelihood of cities adopting these norms and voluntary standards, thereby expanding the reach of transnational climate change governance.

Keohane and Victor further argue that regime complexes may be more adaptable than regimes and can therefore respond more effectively to changing social and political circumstances. International organizations tend to be slow in responding to change; to meet legitimacy concerns, decisions within intergovernmental organizations like the UNFCCC are made through universal voting rules.[26] As it is often impossible to obtain the unanimity of so many countries with diverging interests,

.wbcsd.org/the-european-commission-endorses-wbcsd-set-of-indicators-to-help-cities-advance.aspx (accessed on 20 August 2016). The Asian Cities Climate Change Resilience Network's mission is to build urban climate change resilience in Bangladesh, India, Indonesia, the Philippines, Thailand, and Vietnam through partnerships and collaboration with local and international stakeholders; ACCRN, online: http://acccrn.net/about-acccrn (accessed on 1 August 2016). The International Emissions Trading Association (IETA) is a coalition of companies that supports market-based solutions to climate change. Its members include Bank of America Merrill Lynch, BP, Rio Tinto, and American Electric Power; IETA, online: www.ieta.org/Our-Members (accessed on 1 August 2016).

[24] Young argues that regime interaction is promoted at a national level by domestic policy coordination and at the international level by mutual learning and information sharing between regimes; Margaret A. Young, *Trading Fish, Saving Fish: The Interaction between Regimes in International Law* (Cambridge University Press 2011), pg. 249.

[25] Keohane and Victor, pg. 15.

[26] Article 18 of the UNFCCC states that each member state has one vote in a decision. Further, the Rules of Procedure of the Conference of the Parties state that when parties cannot reach consensus, decisions should be made by majority rule. Tomlinson points out that as these rules have not been formally adopted, the COP usually adopts decisions only when no party explicitly expresses objection;

inertia sets in and renders the regime in question increasingly obsolete. Keohane and Victor argue that regime complexes (and their component parts) may be able to adapt more readily in comparison because they are not hampered in the way that an international regime is. They also argue that a regime complex, with multiple sites of authority, has the conditions to allow the free market of ideas and practices to flourish. The assumption is that the invisible hand will lead to a better distribution of governance functions within the regime complex.

The drawbacks of institutional multiplicity are primarily those of conflicting norms and transaction costs. When voluntary standards are created to satisfy an unmet demand for rule guidance, conflict between potentially overlapping rules is not a major concern given that there are few, if any, existing rules with which new rules may conflict.[27] However, when there are multiple institutions developing and implementing norms, standards, and practices within a regime complex, there remains the possibility of inefficient duplicity and increased transaction costs of complying with different schemes. In the voluntary carbon market, for example, as of December 2008, there were already at least thirteen different programmes offering voluntary GHG accounting standards and methodologies. As the various programmes offer different interpretations and levels of guidance about the accounting standards, this has caused some confusion amongst buyers of carbon credits and project developers about the appropriate standard to use. This situation has led to harmonization efforts to foster more consistency and transparency across voluntary GHG programme accounting standards.[28] Biofuels governance offers another example. Many voluntary certification schemes have emerged to meet consumer demand for sustainably produced biofuels, in addition to the European Union's regulatory requirements governing the sustainability of biofuels that are imported into or produced within EU member states.[29] The certification schemes adopt different methodologies, accounting rules, and definitions, creating confusion and duplicity that result in increased transaction costs for producers. There is currently little impetus for regulatory harmonization, although, as I have argued elsewhere, such

Luke Tomlinson, *Procedural Justice in the United Nations Framework Convention on Climate Change: Negotiating Fairness* (Springer 2015), pg. 157.

[27] Green argues that because of dissent amongst states about the role of emissions trading and thus the utility of greenhouse gases accounting standards, the issue of accounting methodologies was taken off the international agenda and deprived the UNFCCC, the international actor most likely to take on such a role, of the political mandate to do so. This then created an opening for private actors to develop the Greenhouse Gas Protocol (GHG Protocol) when no accounting standards existed; Jessica Green, 'Private Standards in the Climate Regime: The Greenhouse Gas Protocol' (2010) 12 *Business and Politics* Article 3.

[28] For discussion, see Michelle Passero, 'The Voluntary Carbon Market: Its Contributions and Potential Legal and Policy Issues' in David Freestone and Charlotte Streck (eds), *Legal Aspects of Carbon Trading: Kyoto, Copenhagen and Beyond* (Oxford University Press 2009), pgs. 525–527.

[29] European Commission, 'Biofuels', online: https://ec.europa.eu/energy/en/topics/renewable-energy/biofuels (accessed on 1 August 2017).

harmonization would go a long way towards reducing transaction costs.[30] It would render certification affordable for more producers, particularly those operating in developing countries that do not have the economies of scale that multinational conglomerates enjoy.

2.2.2 *Orchestration*

Orchestration has been proposed as a way of managing the disadvantages and maximizing the benefits of institutional fragmentation.[31] Whereas regime complex theory typically treats the co-existence of multiple governance actors with overlapping mandates as a threat to effective governance because of redundancy, conflict, and inconsistency, orchestration theory emphasizes how institutional multiplicity can create gains from the pooling of resources, mutual learning, and specialization. In a pioneering work, orchestration is said to take place when an international organization 'enlists and supports intermediary actors to address target actors in pursuit of [international organization] governance goals. The key to orchestration is that the [international organization] brings third parties into the governance arrangement to act as intermediaries between itself and the targets, rather than trying to govern the targets directly.'[32]

In their alternative definition, Hale and Roger emphasize the relationship between orchestration and transnational governance. They define orchestration to be 'a process whereby states or intergovernmental organizations initiate, guide, broaden, and strengthen transnational governance by non-state and/or sub-state actors'. The transnational governance literature, which focuses on private certification schemes such as the Marine Stewardship Council and standard-setting bodies such as the International Accounting Standards Board (IASB), tends to treat these entities as stand-alone governance actors and pays little attention to their institutional context. Orchestration theory, in contrast, places the analytical focus on the wider relational context in which governance actors operate. It highlights how international organizations shape the capacities and agendas of transnational governance actors. This analytical viewpoint is adopted in this book, particularly in the discussion of the interaction between C40 and the World Bank in Chapter 5.

The intermediaries in orchestration governance are often NGOs but may also include business organizations and transgovernmental networks. When sub-state actors – such as elements of national bureaucracies and city mayors – work with their

[30] Jolene Lin, 'The Environmental Regulation of Biofuels: Limits of the Meta-Standard Approach' (2011) 5 *Carbon & Climate Law Review* 34.

[31] Kenneth W. Abbott, 'Strengthening the Transnational Regime Complex for Climate Change' (2014) 3 *Transnational Environmental Law* 57.

[32] Kenneth W. Abbott et al., 'Orchestration: Global governance through Intermediaries' in Kenneth W. Abbott et al. (eds), *International Organizations as Orchestrators* (Cambridge University Press 2015), pg. 4.

counterparts across borders, they form transgovernmental networks.[33] These networks are attractive intermediaries because they are able to bypass the higher levels of national governments. They are also usually composed of policy experts from national regulatory agencies who command technical expertise and control of national bureaucracies.[34] Kenneth Abbott and his colleagues point out that intermediaries voluntarily participate in orchestration because they share the international organization's governance goals and value its material and ideational support.[35] At the same time, when international organizations provide support to intermediaries, this strengthens the governance capacities of the intermediaries while simultaneously providing international organizations a means by which to influence their agendas and activities.[36] Material support would include financial and administrative assistance to strengthen the intermediaries' operational capacity to pursue governance goals. Ideational support would include technical expertise, formal recognition, and endorsement. Endorsement of the intermediary's activities increases its legitimacy and the social authority that it can bring to bear on target actors. An example of endorsement is the support that General Assembly and Security Council resolutions have given to the Kimberley Process certification scheme for blood diamonds.[37] Endorsement can also be legal, as when the WTO accepts that consistency with standards adopted by the Codex Alimentarius Commission satisfies international trade law.[38]

Chapter 5 examines the World Bank's deepening partnership with C40 through the lens of orchestration theory. The bank provides technical capacity-building opportunities, a range of specialized advisory services, and exclusive funding channels to facilitate action by C40 members. C40 was chosen to be an intermediary because it shares the World Bank's governance objective of tackling climate change in cities as part of a broader agenda of sustainable urban development. Further, the climate solutions that C40 espouses are rooted in the norms of liberal environmentalism, which predicate international environmental protection on the promotion and maintenance of a liberal economic order.[39] This commitment to liberal environmentalism aligns well with the underlying philosophy of the World Bank's Strategic Framework that guides the bank's operational response to new

[33] See discussion in Section 2.2.4, 'The Disaggregated State and Transgovernmental Networks'.

[34] Ibid. [35] Abbott et al., pg. 6. [36] Ibid., pg. 14.

[37] For discussion, see Virginia Haufler, 'Orchestrating Peace? Civil War, Conflict Minerals and the United Nations Security Council' in Kenneth W. Abbott et al. (eds), *International Organizations as Orchestrators* (Cambridge University Press 2015).

[38] For discussion, see Manfred Elsig, 'Orchestration on a Tight Leash: State Oversight of the WTO' in Kenneth W. Abbott et al. (eds), *International Organizations as Orchestrators* (Cambridge University Press 2015).

[39] For discussion, see Steven Bernstein, *The Compromise of Liberal Environmentalism* (Columbia University Press 2001).

development challenges posed by climate change.[40] In brief, the World Bank's partnership with C40 is an example of orchestration by an international organization that plays a significant role in the transnational regime complex for climate change by, amongst other things, convening and endorsing an intermediary's urban climate governance efforts.

2.2.3 *Transnational Law*

Philip Jessup famously coined the term 'transnational law' in his Storrs Lectures at Yale University in 1955 as he looked for a concept to capture the legal regulation of actions and events that transcend state boundaries and can accommodate public and private international law.[41] Jessup's new term also encompassed legal relationships amongst individuals, corporations, and international organizations as well as states.[42] Thus, it was as early as the 1950s that scholars began to point out that the traditional state-centric conception of international law inadequately captured the reality of multiple actors interacting and forming legal relationships that operated across state boundaries.

Harold Koh's influential work introduced the notion of the 'transnational legal process' as a predicate for 'transnational law'.[43] Koh defines transnational legal process as 'the theory and practice of how public and private actors – nation states, international organizations, multinational enterprises, non-governmental organizations, and private individuals – interact in a variety of public and private, domestic and international fora to make, interpret, enforce, and ultimately, internalize rules of transnational law'.[44] According to Koh, there are four distinctive features of the transnational legal process. First, it 'breaks down two traditional dichotomies that have historically dominated the study of international law: between domestic and international, public and private'.[45] Secondly, the transnational legal process is not state-centric in nature: 'the actors in this process are not just, or even primarily,

[40] Under the Strategic Framework, the World Bank Group commits to provide new adaptation financing to vulnerable countries, share lessons to improve the monitoring of climate-related finance, promote the development of carbon markets, and pilot new initiatives to support the development and dissemination of new energy technologies. The key assumptions that underpin this framework are (1) that the World Bank will be able to 'maintain the effectiveness of its core mission of supporting growth and overcoming poverty while recognizing the added costs and risks of climate change' and (2) that the basic tenets of a market economy, economic growth, and environmental protection are all compatible. See *Development and Climate Change: A Strategic Framework for the World Bank Group* (World Bank, Washington DC, 2008), pg. xii, online: http://documents.worldbank.org/curated/en/2012/05/16459433/development-climate-change-strategic-framework-world-bank-group (accessed on 2 April 2016).

[41] Philip Jessup, *Transnational Law* (Yale University Press 1956), pg. 1. [42] Ibid.

[43] Maya Steinitz, 'Transnational Legal Process Theories' in Cesare Romano, Karen J. Alter, and Yuval Shany (eds), *The Oxford Handbook of International Adjudication* (Oxford University Press 2014), pg. 341.

[44] Harold Hongju Koh, 'Transnational Legal Process' (1996) 75 *Nebraska Law Review* 181, pg. 184.

[45] Ibid.

nation-states, but include nonstate actors as well'.[46] Thirdly, the process is dynamic. 'Transnational law transforms, mutates, and percolates up and down, from the public to the private, from the domestic to the international level and back down again.'[47] Finally, it is a normative process. The concept of the transnational legal process 'focuses not simply upon how international interaction among transnational actors shapes law, but also on how law shapes and guides future interactions: in short, how law influences why nations obey'.[48] The crux of Koh's theory of the transnational legal process is, in fact, internalization through interaction.[49] States and non-state actors obey international law as a result of repeated interaction with other actors in the international realm. 'As transnational actors interact, they create patterns of behavior and generate norms of external conduct which they in turn internalize.'[50] Through an iterative process of interaction and internalization, 'international law acquires its "stickiness"'.[51]

Building from Koh's theory of the transnational legal process, Gregory Shaffer, alongside other scholars, has developed an empirical, socio-legal framework to analyze the effects of transnational law on national legal systems and to explain the factors determining the extent, location, and limits of transnationally induced legal change.[52] Shaffer defines transnational law to be 'law in which transnational actors, be they institutions or networks of public or private actors, play a role in constructing and diffusing legal norms'.[53] Legal norms are, in turn, defined as 'norms that lay out behavioral prescriptions issued by an authoritative source that take written form, whether or not formally binding or backed by a dispute settlement or other enforcement system'.[54] The source of the legal norm may be an international treaty, a private certification scheme, or a foreign legal model promoted by transnational actors. In an article for the inaugural issue of the journal *Transnational Environmental Law*, Gregory Shaffer and Daniel Bodansky wrote about the unilateral nature of some transnational legal processes.[55] Such unilateralism usually occurs when states have little choice but to adopt the regulatory standards of dominant market actors such as the United States and European Union. The regulation of chemicals is a case in point. The EU passed a regulation in 2006 known as REACH, which created new and more stringent requirements for chemicals intended for sale in the EU.[56] This created pressure on exporting states to

46 Ibid. 47 Ibid. 48 Ibid. 49 Steinitz, pg. 345. 50 Koh, pg. 204. 51 Ibid.

52 Gregory C. Shaffer (ed), *Transnational Legal Ordering and State Change* (Cambridge University Press 2014).

53 Gregory C. Shaffer, 'Transnational Legal Process and State Change' (2012) 37 *Law and Social Inquiry* 229, pg. 235.

54 Ibid., pg. 234.

55 Gregory C. Shaffer and Daniel Bodansky, 'Transnationalism, Unilateralism and International Law' (2012) 1 *Transnational Environmental Law* 31.

56 Regulation No. 1907/2006 of the European Parliament and of the Council on the Registration, Evaluation, Authorisation and Restriction of Chemicals (REACH).

improve their regulation and production methods, and ratcheted up chemicals regulation beyond the EU without any international treaty.[57]

This book draws on these theories of the transnational legal process and transnational law to conceptualize the role of cities as transnational actors involved in the creation and diffusion of norms pertaining to climate mitigation and adaptation. Chapter 4, for example, describes what five global cities are doing to tackle climate change. It can be viewed as simply an account of local climate action. However, from the transnational law perspective, local climate action is part of an iterative process of creating and implementing legal norms that have transnational significance. Local climate action is part of the dynamic process whereby legal norms percolate up to the international level and down to the city level, and between public and private spheres. Chapter 5 is devoted to examining the C40 network as the horizontal architecture through which cities create and diffuse norms, practices, and voluntary standards, and Chapter 6 explores an important aspect of transnational law that tends to be underappreciated in the literature: the interaction between transnational law and public international law that finds its source in treaties.

2.2.4 *The Disaggregated State and Transgovernmental Networks*

In *A New World Order*, Anne-Marie Slaughter describes a world in which global governance in a wide range of areas such as food safety and human trafficking is conducted by national government officials who cooperate with one another through cross-border networks.[58] Operating across national borders, these networks are transnational in nature and are mostly made up of government officials.[59] Slaughter uses the term 'transgovernmental' to capture these two characteristics of the networks.[60] She argues in *A New World Order* that transgovernmental networks, unlike formal international organizations which are often paralyzed by politics, have the expertise, flexibility, and inclusiveness to solve global collective problems.[61] She further argues that, once a transgovernmental network adopts a set of standards or

[57] For discussion, see Joanne Scott, 'From Brussels with Love: The Transatlantic Travels of European Law and the Chemistry of Regulatory Attraction' (2009) 57 *American Journal of Comparative Law* 897; Yoshiko Naiki, 'Assessing Policy Reach: Japan's Chemical Policy Reform in Response to the EU's REACH Regulation' (2010) 22 *Journal of Environmental Law* 171.

[58] Anne-Marie Slaughter, *A New World Order* (Princeton University Press 2005).

[59] For example, the Basel Committee of Banking Supervision (BCBS) is made up of the central bank governors from twenty-seven countries and the European Central Bank; Basel Committee Membership, online: www.bis.org/bcbs/ (accessed on 8 July 2016). Another example is the Asian Judges Network on Environment (AJNE), which seeks to facilitate judicial capacity building through sharing information and experience; AJNE, online: www.asianjudges.org (accessed on 8 July 2016).

[60] Slaughter's work on transgovernmental networks has its intellectual roots in the 'transgovernmental relations' approach pioneered by political scientists Robert Keohane and Joseph Nye in the 1970s: Robert O. Keohane and Joseph S. Nye, *Power and Interdependence: World Politics in Transition* (TBS The Book Service Ltd 1977); Robert O. Keohane and Joseph S. Nye, 'Transgovernmental Relations and International Organizations' (1974) 27 *World Politics* 39.

[61] Slaughter, pg. 167.

rules, the domestic implementation by national regulators lend these rules 'hard power' and make them effective.[62] Moreover, regulators that participate in a transgovernmental network are usually government officials who are ultimately accountable to their domestic constituencies. Hence, in terms of accountability and legitimacy, transgovernmental networks are preferable to 'amorphous "global policy networks" . . . in which it is never clear who is exercising power on behalf of whom'.[63]

Slaughter uses the terms 'horizontal' and 'vertical' to connote whether a particular network only connects actors at a single level of governance (for example, at the national level) – in which case the network is horizontal – or does so vertically, involving actors at different governance levels (for example, at the levels of the city, the state, and supranational organizations).[64] Horizontal transgovernmental networks may be categorized according to their main function (information sharing, enforcement, and harmonization) although there is often functional overlap in these networks. Information networks, as the name suggests, facilitate the exchange of information and experience.[65] Further, these networks often actively collect and distil information into a code of 'best practices' for achieving a desired regulatory objective. Enforcement networks are most commonly established among law enforcers such as police officers and customs authorities. The focus of enforcement networks is on enhancing cooperation among domestic regulators to enforce national laws in the face of transboundary illicit activity such as illegal wildlife trafficking.[66] Finally, harmonization networks tend to arise when regulators work together to seek convergence of their regulatory policies and standards (such as product safety standards). Trade agreements such as the WTO and the North American Free Trade Agreement (NAFTA) often

[62] Slaughter, pgs. 168–169. For discussion of how the EU, advertently and sometimes inadvertently, institutionalizes and 'hardens' voluntary standards and informal best practices by embedding them in domestic law, thereby shaping the dynamics of transnational norm diffusion processes, see Abraham Newman and David Bach, 'The European Union as Hardening Agent: Soft Law and the Diffusion of Global Financial Regulation' (2014) 21 *Journal of European Public Policy* 430.

[63] Slaughter, pg. 4.

[64] An example of a vertical network exists between national courts and the European Court of Justice in the EU. For discussion of trans-judicial networks, see Elaine Mak, *Judicial Decision-Making in a Globalised World: A Comparative Analysis of the Changing Practices of Western Highest Courts* (1st edn, Hart Publishing 2013), pgs. 83–95.

[65] An example is the Pharmaceutical Inspection Co-operation Scheme, which provides for the exchange of information amongst health authorities on good manufacturing practices and also undertakes efforts to harmonize technical standards and procedures regarding inspection of the manufacture of medicinal products; Pharmaceutical Inspection Co-operation Scheme, online: www.picscheme.org /role.php (accessed on 8 April 2016).

[66] Slaughter, pgs. 55–58. The ASEAN Wildlife Enforcement Network, for example, seeks to promote cooperation and collaboration between national law enforcement agencies, customs agencies, the police, and prosecutors to enforce wildlife trafficking laws; ASEAN Wildlife Enforcement Network, online: http://environment.asean.org/the-asean-wildlife-enforcement-network-asean-wen/ (accessed on 8 April 2016). For discussion about the role of enforcement networks in environmental governance, see Michael Faure, Peter De Smedt, and An Stas (eds), *Environmental Enforcement Networks: Concepts, Implementation and Effectiveness* (Edward Elgar Publishing 2015).

require harmonization, 'resulting in harmonization networks of countries moving toward a single standard'.[67]

In Slaughter's vision, these transgovernmental networks will not replace the existing infrastructure of international institutions but rather will complement and strengthen them. In some cases, transgovernmental networks emerge because states fail to reach international consensus and therefore are unable to develop common rules and coordinate responses to challenges posed by globalization. In competition law, for example, Cheng suggests that the abandonment of the Singapore agenda in the WTO (which included competition policy) led the international competition community to focus on voluntary convergence through transgovernmental networks.[68] In this way, transngovernmental networks can serve to fill governance gaps and, through cooperation amongst government officials, create the degree of consensus that could lead to a treaty-based regulatory regime if that is a desired outcome.

Transgovernmental networks may also be a more effective governance mechanism, compared to a treaty-based regulatory framework, because they harness the benefits of the network form – namely speed and the lack of bureaucratic formality. Simmons argues that, in the case of international finance, its nature necessitates that governance be carried out mostly through transgovernmental networks: 'Formal, protracted negotiations would be rapidly overtaken by technological change, financial innovation, and other market developments.'[69]

Underlying this vision of a world order made up of networks is a major shift in thinking about the international system. Slaughter argues that the state is *disaggregating* and its component governmental institutions – regulators, judges, and even legislators – are reaching out to their foreign counterparts and creating transnational networks.[70] Instead of thinking of the international system as comprising states as 'unitary entities like billiard balls or black

[67] Slaughter, pg. 59. There is the general view that regulatory convergence is beneficial for trade because it reduces the costs of complying with differing product standards. However, for a critical view of how harmonization lowers regulatory standards concerning drug safety and makes the state more vulnerable to capture by commercial interests, see John Abraham and Tim Reed, 'Trading Risks for Markets: The International Harmonisation of Pharmaceuticals Regulation' (2001) 3 *Health, Risk & Society* 113.

[68] Thomas K. Cheng, 'Convergence and Its Discontents: A Reconsideration of the Merits of Convergence of Global Competition Law' (2012) 12 *Chicago Journal of International Law* 433, pg. 435. The International Competition Network (ICN) is widely credited for being the most important agent of regulatory convergence in competition policy in the past decade and is said to '[exert] its greatest influence through the recommended practices and other work products produced by its working groups'; Cheng, pg. 443.

[69] Beth A. Simmons, 'The International Politics of Harmonization: The Case of Capital Market Regulation' (2001) 55 *International Organization* 589, pg. 592.

[70] Also see Kal Raustiala, 'The Architecture of International Cooperation: Transgovernmental Networks and the Future of International Law' (2002) 43 *Virginia Journal of International Law* 1: Instead of disappearing, the state is 'disaggregating for purposes of cooperation: domestic officials are reaching out to their foreign counterparts regularly and directly through networks, rather than through state-to-state negotiation of the kind that dominated 20th century cooperation'; pg. 10.

boxes',[71] recognizing the disaggregation of states prompts us to think of the international system as a more complex landscape made up of international organizations, states (in their unitary form), networks made up of parts of a state (for example, the judiciary), and private actors. In Slaughter's view, states can be disaggregated but also act as completely unitary actors when necessary (for example, when deciding to go to war).[72]

Slaughter argues that transgovernmental networks are 'a key feature of world order in the twenty-first century, but they are underappreciated, undersupported, and underused to address the central problems of global governance'.[73] She makes the claim that these networks should be 'embraced' as 'the architecture of a new world order'.[74] However, it should be noted that other commentators have expressed reservations about the ways in which transgovernmental networks signal a shift towards global governance by experts acting outside the constraints of domestic political structures and the normal foreign affairs processes.[75] Philip Alston argues that the formation of these networks 'suggest[s] a move away from arenas of relative transparency into the back room' leading to 'the bypassing of the national political arenas to which the United States and other proponents of the importance of healthy democratic institutions attach so much importance'.[76] Alston also raises the interesting point that Slaughter's analysis brings to the fore two key issues: 'They are: (i) what is the nature of the global agenda in a globalized world? and (ii) who sets and implements that agenda?'[77] In Sol Picciotto's view, transgovernmental networks result in the dispersal of politics into functional arenas and '[appear] to allow particular issues to be regulated in a depoliticized, technocratic manner, by mangers or professionals who are directly accountable to their "customers"'.[78] Finally, Stephen Toope argues that networks are no different from regimes in that they are also sites of power that are capable of exclusion and inequality. Operating outside

[71] Slaughter, pg. 5. [72] Ibid., pg. 19. [73] Ibid., pg. 1. [74] Ibid., pg. 213.

[75] In addition to the views expressed in the remainder of this paragraph, also see David Kennedy, 'The Politics of the Invisible College: International Governance and the Politics of Expertise' (2001) 5 *European Human Rights Law Review* 463; Pierre-Hugues Verdier, 'Transnational Regulatory Networks and Their Limits' (2009) 34 *Yale Journal of International Law* 113. Verdier argues that the existing scholarship has tended to downplay conflicts of interest within transgovernmental networks. He points out that most of the scholarship provides highly detailed and optimistic accounts of successful transgovernmental networks but offers few accounts of networks that have failed in their mission. There is also limited discussion of the use of transgovernmental networks by powerful states to impose their preferred standards on less powerful states or the failure of transgovernmental networks to prevent non-adherence to their standards. After all, transgovernmental networks are intrinsically hampered from effectively addressing enforcement problems because they do not have monitoring and enforcement capabilities; pgs. 121–122.

[76] Philip Alston, 'The Myopia of the Handmaidens: International Lawyers and Globalization' (1997) 8 *European Journal of International Law* 435, pg. 441.

[77] Ibid., pg. 439.

[78] Sol Picciotto, 'Fragmented States and International Rules of Law' (1997) 6 *Social and Legal Studies* 259, pg. 273.

the realm of public scrutiny, these networks may be even less accountable than some states in Toope's view.[79]

In response to these critiques, Slaughter has argued that accountability of transgovernmental networks has to be fostered at the national and global levels.[80] Further, in order to develop or adapt accountability mechanisms, she argues that distinctions must be drawn between networks that operate in existing international organizations, within the framework of an executive agreement, or outside any preexisting formal framework.[81] Finally, she proposes a set of global norms that can be used by members of transgovernmental networks to govern their relations with one another.[82] These norms include global deliberative equality, legitimate difference, positive comity, checks and balances, and subsidiarity.[83] They are meant to set the ground rules, which may be redundant when a network is mainly engaged in information exchange but which will be necessary when networks engage in harmonization and enforcement activities.[84]

Viewing the international system through the prism of disaggregated states brings to the foreground features of the climate change governance landscape that remain hidden from a traditional statist perspective. The most prominent revelation is the sheer amount of sub-national governance activity that is taking place. For example, city-level government officials and mayors have been reaching out to their foreign counterparts and creating information networks that are beginning to play a role in the generation and diffusion of practices and norms. Using the literature on the disaggregated state and transgovernmental networks as a point of departure, this book conceptualizes transnational city networks as a variant of transgovernmental networks. Unlike transgovernmental networks, transnational city networks constitute a form of disaggregation along the vertical levels of government. However, like transgovernmental networks, transnational city networks have emerged as a tool of governance. Cities and their governments are using these networks to supplement governance gaps in the transnational climate change regime complex and push for more concerted climate mitigation

79 Stephen Toope, 'Emerging Patterns of Governance and International Law' in Michael Byers (ed), *The Role of Law in International Politics: Essays in International Relations and International Law* (Oxford University Press 2001), pgs. 96–97.

80 Anne-Marie Slaughter, 'The Accountability of Government Networks' (2001) 8 *Indiana Journal of Global Legal Studies* 346. See Verdier, who questions whether domestic political constraints operate as intended in the context of international regulatory cooperation. He further argues that if domestic political constraints were to act as accountability mechanisms for individual regulators involved in a transgovernmental network, then these regulators would be bound to domestic interests rather than to some shared sense of regulatory common good. Domestic interests are also likely to cause disagreement on issues such as which rules to adopt, and attempts to dilute or resist standards or avoid compliance; pgs. 126–130.

81 Slaughter, ibid., pg. 349.

82 Anne-Marie Slaughter, 'Disaggregated Sovereignty: Towards the Public Accountability of Global Government Networks' (2004) 39(2) *Government and Opposition* 159.

83 Ibid., pgs. 174–186.

84 Ibid., pg. 175.

and adaptation efforts at the local level. They are also using networks to develop and disseminate standards and practices that are specifically tailored to meet the needs of the city and its inhabitants in the face of climate change. The critiques of transgovernmental networks serve as a basis for reflecting upon the limitations of transnational city networks and issues of accountability and transparency that may arise when these networks grow in their reach and normative influence. Finally, the questions that Philip Alston has posed concerning the content of the global governance agenda and who sets it serve as a departure point for the enquiry in Chapter 3 into the urban sustainable development agenda and the role of the World Bank in shaping and implementing it.

2.2.5 *The Global City*

Global city theory developed in the late 1980s as scholars sought to understand how the world economy was being transformed. Drawing upon traditional urban theory (which examined how cities related to the higher levels of government within domestic political systems) and world systems theory (which analyzed how states were bound in unequal structural core–periphery relationships using a neo-Marxist framework), John Friedmann published his 'World City Hypobook' in 1986.[85] Friedmann understood the city in economic terms – cities were 'basing points' and 'organisational nodes' of the rapidly changing geography of global capitalism in the twentieth century – and sought to analyze the 'spatial organization of the new international division of labour'.[86] This 'new division of labour', Brenner explains, is largely the result of the massive expansion in the role of transnational corporations in the production and exchange of commodities globally since the late 1960s.[87] 'Whereas the old international division of labour was based upon raw materials production in the periphery and industrial manufacturing in the core, the [new division of labour] has entailed the relocation of manufacturing industries to semi-peripheral and peripheral states in search of inexpensive sources of labor power.'[88] Meanwhile, business services began to concentrate themselves in urban centres, and these 'upper tier cities' evolved to become major nodes of financial planning and corporate decision-making, and therefore the 'central basing points for the worldwide activities of [transnational corporations]'.[89] Curtis argues that 'Friedmann's key contribution was to place cities back on the agenda of international political economy, showing how the internal life of cities . . . could only be understood by reference to their connections at the international level and the functions that they fulfill for the global economy.

[85] John Friedmann, 'The World City Hypobook' (1986) 17 *Development and Change* 69. [86] Ibid.
[87] Neil Brenner, 'Global Cities, Glocal States: Global City Formation and State Territorial Restructuring in Contemporary Europe' (1998) 5 *Review of International Political Economy* 1, pg. 5.
[88] Ibid. [89] Ibid.

At the same time, the global economy could only be properly understood by reference to the role that certain cities play within it.'[90]

The 1990s witnessed the acceleration of economic globalization and advances in information and communication technologies. Against this background, Saskia Sassen produced 'The Global City', which has become a seminal work on the connection between certain key metropolises and the broader processes of globalization.[91] Sassen argues that, alongside the spatial dispersal of economic activities and the increasing digitizing of economic activities such as banking, was the spatial concentration of highly specialized professional firms and top-level management. This process of spatial concentration, and consequently the concentration of material facilities, occurred in cities. Sassen explains that '[a]t the heart of this deep structural trend is the fact that even the most material economic sectors (mines, factories, transport systems, hospitals) today are buying more insurance, accounting, legal, financial, consulting, software programming, and other such services for firms. These so-called intermediate services tend to be produced in cities, no matter the nonurban location of the mine or the steel plant that is being serviced.'[92]

Global cities stand out in comparison to other cities because they are able to handle the more complex needs of firms that have global supply chains or have operations in multiple jurisdictions. Sassen posits that there is no such thing as a single global city.[93] A global city can only exist alongside other global cities, connected by networks that crisscross the world. Each network is likely to include different cities to serve various global economy needs and provide specialized services to multinational corporations. For example, the circuit for the global coffee trade (including the trading of coffee as a commodity, selling it as a consumer product, and the trading of coffee futures, for example) involves cities different from those in the circuit for legal services. Sassen points out that the proliferation of such circuits is not solely driven by economic forces. Non-profit organizations working for labour rights and environmental protection also create and develop global city networks to forge links with their counterparts elsewhere.[94]

As Sassen explains in her later work, her use of the term 'global city' was a conscious decision to mark a departure from the 'older historical term *world city*'.[95] For Sassen, 'global city' is not a descriptive term for a distinctive urban form: 'It is an analytic construct that allows one to detect the global as it is filtered through the specifics of

[90] Simon Curtis, 'Global Cities and the Transformation of the International System' (2011) 37 *Review of International Studies* 1923, pg. 1930.

[91] Saskia Sassen, *The Global City: New York, London, Tokyo* (Princeton University Press 1991).

[92] Saskia Sassen, *Cities in a World Economy* (4th edn, Sage Publications 2012), pg. 110.

[93] Ibid., pgs. 7, 111. Also see Saskia Sassen, 'The Global City: Introducing a Concept' (2005) XI *Brown Journal of World Affairs* 27, pg. 34.

[94] Sassen, *Cities in a World Economy*, pg. 113.

[95] Saskia Sassen, 'Foreword' in Mark M. Amen, Kevin Archer, and Martin M. Bosman (eds), *Relocating Global Cities: From the Center to the Margins* (Rowman & Littlefield Publishers 2006), pg. ix.

a place, its institutional orders, and its sociospatial fragmentations.'[96] In other words, given that the processes of globalization create global cities and global cities, in turn, are territorial spaces in which the processes of globalization unfold and affect local communities, the global city is a theoretical prism through which the connections between cities and globalization can be studied.

Most global city researchers focus on the global economy.[97] However, there are two strands of scholarship that depart from this focus on the provision of services to facilitate global capital movements and meeting the corporate needs of multinational corporations. One strand seeks to unravel the 'dark side' of the global city phenomenon; this includes the extreme social inequality and marginalization of the poor within global cities as well as the detrimental impact that a global city can have on the rest of the country in which it is located. A good example of the latter would be the impact of London's 'global city-ness' on the rest of the United Kingdom.[98] The other strand of scholarship makes the claim that the global city has become an actor in world affairs.[99] Acuto, for instance, seeks to highlight the agency of global cities: 'Admitting that certain cities (if not all cities) perform functions, are capable of innovation and retain degrees of control, implies, in my view, a logical corollary: global cities, due to their presence as loci of purposive action within the global system and as articulators of global flows, are not only places but also participants in world affairs.'[100] Acuto further points out that this is an important analytical step to take because recognizing the capacity of global cities to act allows us to then take the further step of considering their potential role in solving global governance challenges.[101] Arguing that the time has come to re-examine what it means to be a global city and to govern a global city as 'many local governments of Global Cities around the world . . . have broken the institutional chains of municipal politics and have become global actors', Ljungkvist's *The Global City 2.0* offers a rich theoretical examination of the political agency of cities in relation to foreign and security affairs that explains their increasing participation in global politics.[102]

[96] Ibid., pg. x.

[97] For an overview of this body of contemporary scholarship, see Michele Acuto, 'Finding the Global City: An Analytical Journey through the "Invisible College"' (2011) 48 *Urban Studies* 2953, pgs. 2964–2967.

[98] See, for example, Doreen Massey, *World City* (Polity Press 2007).

[99] See, for example, Kent E. Calder and Mariko de Freytas, 'Global Political Cities as Actors in Twenty-First Century International Affairs' (2009) 29 *SAIS Review of International Affairs* 79; Michele Acuto, *Global Cities, Governance and Diplomacy: The Urban Link* (Routledge 2013); Yishai Blank, 'Localism in the New Global Legal Order' (2006) 47 *Harvard International Law Journal* 263; Janne E. Nijman, 'Renaissance of the City as Global Actor: The Role of Foreign Policy and International Law Practices in the Construction of Cities as Global Actors' in Gunther Hellmann, Andreas Fahrmeir, and Milos Vec (eds), *The Transformation of Foreign Policy: Drawing and Managing Boundaries from Antiquity to the Present* (Oxford University Press 2016); Kristin Ljungkvist, *Global City 2.0: From Strategic Site to Global Actor* (Routledge 2016).

[100] Acuto, 'Finding the Global City: An Analytical Journey through the "Invisible College"', pg. 2967.

[101] Ibid., pg. 2968. [102] Ljungkvist.

This book taps into the large and broad research programme on global cities as a point of departure for conceptualizing cities as actors and the involvement of cities in the processes of globalization. In particular, the global city scholarship provides valuable insights into how certain cities are capable of participating in global climate governance processes because of their ability to forge transnational, networked relations and why global cities want to be actors in world affairs (including the governance of climate change). This has created the basic foundation for the exploration of cities as lawmaking actors in the context of transnational climate change governance.

2.3 CONCLUSION

This chapter presents the overarching theoretical framework for this book. In the spirit of analytical eclecticism, it draws on diverse theoretical perspectives to create a framework for examining the emergence of cities as participants in transnational climate change law and governance. Chapter 3 provides an account of the rise of cities as participants in world affairs and the practices of international organizations like UN-Habitat that augment the increasingly popular idea that in the face of grave global governance challenges – such as climate change, pandemics, and human trafficking – cities can offer pragmatic and effective solutions while states appear to be dysfunctional and ineffectual.[103] These practices and their underlying assumptions about the positive role that cities can play in world affairs encourage cities to exercise their agency in global governance, as we will see in Chapter 3.

[103] See, for example, Benjamin Barber, *If Mayors Ruled the World: Dysfunctional Nations, Rising Cities* (Yale University Press 2014); Ljungkvist; UN-Habitat, *Cities in A Globalizing World – Global Report on Human Settlements 2001* (Earthscan 2001).

3

The Rise of the City in International Affairs

In short, global cities are increasingly driving world affairs – economically, politically, socially and culturally. They are no longer just places to live in. They have emerged as leading actors on the global stage.[1]

Given that by 2050, eight out of ten people in the world will be living in cities, 'the battle for a more sustainable future will be won or lost in cities.' [2]

3.1 INTRODUCTION

Cities have existed since prehistoric times. Over the past 2,000 years, a few cities have emerged as international political centres. These were mostly imperial cities – such as Thebes, Babylon, Persepolis, Rome, and Constantinople – which were 'capitals of great empires and their conquered territories'.[3] In China, the Spring and Autumn period (771–481 BC) was the age of the city state; the area surrounding the Yellow River was divided into hundreds of states, most of which consisted of a single city and its hinterland.[4] In more recent history, cities including Vienna, St Petersburg, Amsterdam, and London also asserted significant social and political influence on the international stage. These cities wielded a significant amount of power and influence because of the strength and wealth of the empires over which they ruled.[5]

[1] Ivo Daalder, 'A New Global Order of Cities' *Financial Times* (26 May 2015).

[2] World Urban Campaign, *Manifesto for Cities: The Urban Future We Want* (2012).

[3] Siddhartha Sen, 'Imperial Cities' in Roger W. Caves (ed), *Encyclopedia of the City* (Routledge 2005), pgs. 250–251.

[4] Mark Edward Lewis, 'The City-State in Spring-and-Autumn China' in Mogens Herman Hansen (ed), *A Comparative Study of Thirty City-State Cultures: An Investigation*, Volume 21 of Historisk-filosofiske skrifter (Kgl. Danske Videnskabernes Selskab 2000), pg. 359.

[5] Kent E. Calder and Mariko de Freytas, 'Global Political Cities as Actors in Twenty-First Century International Affairs' (2009) 29 *SAIS Review of International Affairs* 79, pg. 81.

Urbanization in the twentieth-first century differs from the past in its rate and scale. In 1800, when the global population was around a billion people, Beijing was the only city in the world with more than a million inhabitants.[6] In 2014, Tokyo was the world's largest city, with 38 million inhabitants.[7] The UN projects that by 2030, the world will have forty-one megacities with a population of at least 10 million inhabitants each.[8] In 1950, one-third of the world's population lived in cities.[9] This figure is expected to rise to two-thirds by 2050.[10] It should also be noted that there is invariably a close connection between rural areas and urban metropolises. Rood and his colleagues, for example, have demonstrated that a land area about four times larger than the Netherlands is used to meet the food and fibre needs of this highly urbanized country.[11] It is therefore not an exaggeration to say that we are living in an 'urban age'.[12]

It has been argued that in the urban age, cities, more so than states, 'will be forced into the frontlines by global warming, water insecurity, and other environmental challenges'.[13] Other global governance challenges increasingly have an urban face; terrorism and ethnic violence are just two examples. As cities seek to protect their inhabitants from threats such as terrorism and pandemics, they begin to claim political authority and to develop independent policies in relation to foreign affairs and security. These are issue areas that are traditionally deemed core responsibilities of national governments. Further, by forging transnational partnerships and creating organizations to represent their collective interests at the global level, cities no longer behave like 'passive players in a global game played out among and between national and international actors'.[14] Situated at the conflux of urbanization and globalization, global cities, in particular, are leading the trend of increasingly intense and broad urban participation in international affairs. Global cities are

[6] T. Chandler, *Four Thousand Years of Urban Growth: An Historical Census* (St. David's University Press 1987), pg. 656.

[7] UN, *World Urbanization Prospects: The 2014 Revision, Highlights (ST/ESA/SER.A/352)* (Department of Economic and Social Affairs, Population Division, 2014), online: http://esa.un.org/unpd/wup/Highlights/WUP2014-Highlights.pdf (accessed on 20 April 2016), pg. 1.

[8] Ibid. For a helpful graphical representation of where these cities are and their projected population changes from 2014 to 2030, see Bloomberg Visual Data, 'At Home in a Crowd', online: www.bloomberg.com/infographics/2014–09-09/global-megacities-by-2030.html (accessed on 3 December 2016).

[9] UN, ibid. [10] Ibid.

[11] G. A. Rood et al., *Tracking the Effects of Inhabitants on Biodiversity in the Netherlands and Abroad: An Ecological Footprint Model* (Netherlands Environmental Assessment Agency/RIVM, Publication number 500013005 [Bilthoven, the Netherlands], 2004).

[12] Ricky Burdett and Deyan Sudjic (eds), *The Endless City* (Phaidon Press 2008).

[13] Saskia Sassen, 'A Focus on Cities Takes Us beyond Existing Governance Frameworks' in Joseph Stiglitz and Mary Kaldor (eds), *The Quest for Security: Protection without Protectionism and the Challenge of Global Governance* (Columbia University Press 2013), pg. 244.

[14] Ash Amin and Nigel Thrift, 'Citizens of the World: Seeing the City as a Site of International Influence' (2005) 27 *Harvard International Review* 14, pg. 17.

'[demonstrating] how states and international organizations are no longer the only problem-solving units in world politics'.[15]

However, the role played by cities in international affairs has received very little attention from international law and international relations scholars. From a legal perspective, the city has traditionally been constructed from within a national legal order. As Jerry Frug has argued in his seminal work on American local government law, there have always been two competing notions of the city in co-existence.[16] The bureaucratic model posits the city as a creature of the state, subordinate to the state and only capable of exercising powers that have been delegated to it. The democratic model posits the city as a quasi-sovereign that has limited but real autonomy; the city is a space in which citizens can pursue their collective vision of the 'commonwealth' with minimal state interference.[17] At different points in history and in different places, one of these two notions of the city would gain ascendency. Further, Frug argues, two competing narratives about the city-state relationship have emerged from this tension between the dual notions of the city.[18] One narrative speaks of the role of the city as protector of citizens' interests against the encroachment of oppressive state power, and the other presents the state as the protector of minority interests against the abuse of power by majoritarian rule in cities.[19] As in any system of government that comprises multiple levels, each level of government will occasionally seek to aggregate its power relative to other levels. The management of these power relations, and division of responsibilities between the city and the state, has traditionally been considered a purely domestic matter, not a proper matter of concern for the international community.

Furthermore, cities have traditionally been viewed as 'policy takers' rather than policymakers within the hierarchies of domestic political systems.[20] As for external relations, states have traditionally been reluctant to acknowledge that cities may have a role to play outside the domestic realm. The Dutch national government, for example, explicitly denied in the early 1990s the possibility of local governments possessing competencies on foreign policy.[21] Thus, even though scholars of international law and international relations (except the realists) have long acknowledged that states are not the only actors in the international system, and there is extensive literature on the involvement of actors 'beyond the state' – including individuals, multinational corporations, and religious organizations – in international

[15] Michele Acuto, 'City Leadership in Global Governance' (2013) 19 *Global Governance* 481, pg. 495.
[16] G. E. Frug, 'The City as a Legal Concept' (1980) 93(6) *Harvard Law Review* 1057.
[17] Frug, pgs. 1062–1075. [18] Frug, pgs. 26–53. [19] Ibid.
[20] See Claus Schultze, 'Cities and EU Governance: Policy-Takers or Policy-Makers?' (2003) 13 *Regional and Federal Studies* 121, in which the author argues that cities are increasingly moving from policy 'taker' to maker, signalling a move towards more participative governance in the EU.
[21] Letter of the Minister of Foreign Affairs to the Second Chamber of Parliament, Tweede Kamer, 26 January 1990, Vergaderjaar 1989–1990, 21 300 VII, nr. 20.

lawmaking and world politics, cities have not been acknowledged as having any significant role to play in the international sphere.[22]

This chapter makes the argument that correcting this scholarly oversight is of timely importance in light of contemporary developments. As the empirical account in this chapter will show, cities are increasingly participating in global governance processes and international affairs in various ways. They are engaging in inter-city diplomacy and developing external relations to promote trade and political stability.[23] Cities are also independently implementing international treaties on their own accord, sometimes in opposition to the stance adopted by their states, thereby challenging traditional conceptions of how international law is internalized and implemented domestically.[24] Cities have also formed organizations to represent their interests in international forums and to achieve foreign policy objectives such as nuclear non-proliferation.[25]

The 'internationalization of cities' has also gained traction because of a shift in the development agenda of international organizations, particularly the World Bank and UN-Habitat, towards sustainable urbanization.[26] This shift can be dated to the late 2000s, which heralded the 'Decade of the City, a decade that will be remembered for recognizing cities at the core of growth and human development.'[27] The contemporary global urban agenda has created programmes and mechanisms that link cities directly to international organizations. In this way, national governments are being bypassed. These programmes also channel resources directly to cities, which enhances their position vis-à-vis the state. While the practical effects of these urban programmes and policies are mostly experienced in cities in developing countries and least developed countries, as these cities are usually the recipients of aid and technical assistance from international organizations, the ideological influence of the global urban agenda is indeed global and creates a normative environment that is conducive for cities to play an increasingly visible role in international affairs.

[22] For a sample of this extensive literature, see, for example, Philip Alston (ed), *Non-State Actors and Human Rights (Collected Courses of the Academy of European Law)* (Oxford University Press 2005); Steve Charnovitz, 'Nongovernmental Organizations and International Law' (2006) 100 *American Journal of International Law* 348; Elisa Morgera, *Corporate Accountability in International Environmental Law* (Oxford University Press 2009). On the underappreciation of the role of state courts in international lawmaking, see Janet Koven Levit, 'A Tale of International Law in the Heartland: Torres and the Role of State Courts in Transnational Legal Conversation' (2004) 12 *Tulsa Journal of Comparative and International Law* 163. Exceptions that consider the relationship between cities and international law include Yishai Blank, 'Localism in the New Global Legal Order' (2006) 47 *Harvard International Law Journal* 263; Lesley Wexler, 'The Promise and Limits of Local Human Rights Internationalism' (2009) 37 *Fordham Urban Law Journal* 599; G. E. Frug and David Barron, 'International Local Government Law' (2006) 38 *Urban Lawyer* 1.

[23] See Section 3.2.2. [24] See Section 3.2.1. [25] See Section 3.2.4.

[26] I borrow the term 'internationalization of cities' from Ileana Porras, 'The City and International Law: In Pursuit of Sustainable Development' (2008) 36 *Fordham Urban Law Journal* 537.

[27] World Bank, *Systems of Cities: Harnessing Urbanization for Growth and Poverty Alleviation* (Washington, DC: World Bank 2009), Foreword, pg. 1.

National governments increasingly recognize that cities can play a meaningful role in materializing domestic and global agendas, especially on urban, environmental, and economic issues. Even an authoritarian government of a centralized political system like China's has been encouraging of Chinese cities' global engagement. Through the state-organized China Association of Mayors, Chinese mayors have recently spent time in the United States and Denmark to learn from their counterparts and experts about a range of sustainability issues, including public transportation, protecting open spaces, and energy conservation.[28] In 2014, the mayors of Guangzhou, Los Angeles, and Auckland signed a memorandum of understanding (MOU) to promote economic cooperation.[29] This was followed by the launch of the Tripartite Ports Alliance in 2015, which aims to foster collaboration on investments, technologies, and environmental policies amongst the three cities' port authorities.[30] As of February 2015, 444 Chinese cities have established 2,154 partnerships with cities abroad.[31]

By examining various ways in which cities play a part in global governance processes and international organizations' urban policy, this third chapter provides a backdrop for subsequent analysis of the participation of cities in transnational climate change governance. Section 3.2 provides an overview of four broad categories of activity that exemplify how cities are emerging as participants in global governance processes, usually adopting practices and policy positions that are independent of their states. These four categories are (1) a city implementing international law on its own accord when its national government is reluctant or refuses to do so, (2) city diplomacy, (3) cities developing their independent local and transnational policies and strategies to manage global risks such as terrorism, and (4) cities forming organizations to represent urban interests in international forums and/or to pursue governance objectives such as promoting nuclear non-proliferation. Section 3.3 examines the normative policy platform and associated practices of international organizations, particularly UN-Habitat and the World Bank, that constitute the contemporary global urban agenda. Section 3.4 concludes by providing an overview of the key ideas advanced in this chapter.

[28] Wen Huang, 'Delegation of Chinese Leaders Study Sustainable Urbanization' *Granham Global Initiatives* (University of Chicago, 7 August 2015), online: https://grahamglobal.uchicago.edu/news/delegation-chinese-leaders-study-sustainable-urbanization; Grundfos, *Grundfos to Train Chinese Mayors* (27 November 2014), online: www.grundfos.com/about-us/news-and-press/news/grundfos-to-train-chinese-mayors.html (both webpages accessed on 1 May 2016).

[29] Xinhua News Agency, *Leaders from China, New Zealand, US Seek Common Prosperity at Tripartite Meeting* (16 May 2016), online: http://china.org.cn/world/2016–05/16/content_38460907.htm (accessed on 1 May 2016).

[30] Port of Los Angeles, *Ports of Los Angeles, Auckland and Guangzhou Establish Tripartite Ports Alliance* (11 June 2015), online: www.portoflosangeles.org/newsroom/2015_releases/news_061115_tripartite.asp (accessed on 1 May 2016).

[31] Danyang Wang, 'Analyzing China's "City Diplomacy"' *Yicai* (21 May 2015), online: www.yicai.com/news/4620833.html (accessed on 1 May 2016).

3.2 URBAN PARTICIPATION IN INTERNATIONAL LEGAL AND POLITICAL PROCESSES

3.2.1 *Cities Implementing International Law*

According to most orthodox accounts of international law, when a state signs and ratifies a treaty, it expresses its commitment to upholding an international norm or set of norms. Subsequently, international law depends on domestic legal systems for implementation.[32] The act of incorporating international law into its domestic legal and political structures – through executive action, legislation, and judicial decisions – is an expression of a state's internalization of international law. Internalization by a state trickles from the national level down to the state level and further down to cities when these sub-national entities implement the law. Thus, according to orthodox accounts, sub-state units have a circumscribed role to play in the internalization of international law norms which is limited to implementing national laws and policies. However, the internalization of international law can occur in a less rigid, bottom-up manner that breaks the monopoly of nations on international lawmaking processes. When a state is reluctant or refuses to ratify a treaty, cities have, on occasion, served as points of entry for international norms into a state when they introduce policies and legislation to implement international law.

A city may choose to implement an international treaty because it wishes to express its cosmopolitan identity by signalling its membership of the global community.[33] A city may also already have a strong domestic commitment to the subject matter of the international treaty and choose to implement the international treaty so that it can serve as a focal point and provide widely recognized indicators for guidance.[34] For example, a city that is committed to addressing climate change may choose to implement the Paris Agreement so that it serves as a focal point for the city's actions and policies. The two-degree Celsius target and the Nationally Determined Commitments (NDCs) (known as the Intended Nationally Determined Commitments before the Paris Agreement came into force on 4 November 2016) provide a framework of indicators for the city to measure its

[32] 'At the heart of any chapter on international law and national law is always an explanation of the two theories of monism and dualism'; Rosalyn Higgins, *Problems and Process: International Law and How We Use It* (Oxford University Press 1994), pgs. 205–206. Monism is a theory positing that all law, including international law, is part of a single universal legal order. Therefore, international law exists alongside the various branches of domestic law (for example, labour law and company law). Dualism envisages international law and domestic law as two distinct spheres. In order for international law to become part of a domestic legal system, it must be 'transformed' into domestic legislation. For discussion, see Peter Malanczuk, *Akehurst's Modern Introduction to International Law* (7th revised edn, Routledge 1997), chapter 4.

[33] Dan Koon-hong Chan, 'City Diplomacy and "Glocal" Governance: Revitalizing Cosmopolitan Democracy' (2016) 29 *Innovation: The European Journal of Social Science Research*, DOI: 10.1080/13511610.2016.1157684.

[34] Also see discussion that follows on the Convention on the Elimination of All Forms of Discrimination Against Women.

performance. The city can choose to exceed the NDC of its state, it can endorse the temperature target in its climate policies and action plans, or it can pledge the equivalent of an NDC and report its emissions in the same way that member states will be doing pursuant to the Paris Agreement; the Compact of Mayors provides cities the platform for doing so.[35] A city may elect to implement a treaty or base its policies on international legal norms to express its criticism of national policy.[36] The UK city of York offers an interesting example. At a time of strong resistance to the Human Rights Act in the country, the city intentionally chose to frame its local policies in terms of international human rights.[37] York also welcomed refugees in order to mitigate increasingly severe national migration policies.[38] Behind various possible motivations to implement international law, the extent to which an international issue becomes part of a city's agenda will depend heavily on the strength and influence of constituency demands, interest group advocacy and lobbying activities, and local officials' responsiveness.[39]

At this juncture, I give three examples of cities implementing international treaties on their own accord and therefore 'bringing international law home' despite non-ratification by their national governments.[40] The United States had signed the Kyoto Protocol on 12 November 1998.[41] However, one year after the White House came under the control of the Republican Party following the presidential elections in 2000, the Bush administration withdrew the US signature from the Kyoto Protocol on the basis of scientific uncertainty and unfairness.[42] However, many local officials in the United States did not share President Bush's opposition to the Kyoto Protocol. On the day that the Kyoto Protocol became law for the countries that had signed and ratified it, the mayor of Seattle, Greg Nickels, launched the US Conference of Mayors Climate Protection Agreement whereby signatory cities pledged to surpass the GHG reduction target suggested for the United States in the Kyoto Protocol (i.e. 7 per cent reduction from 1990 levels by 2012) as well as urge the United States

[35] The Compact of Mayors is discussed in detail in Chapter 5.

[36] Lesley Wexler, 'Take the Long Way Home: Sub-Federal Integration of Unratified and Non-Self-Executing Treaty Law' (2006) 28 *Michigan Journal of International Law* 1.

[37] Esther van den Berg and Barbara Oomen, 'Towards a Decentralisation of Human Rights: The Rise of Human Rights Cities' in Thijs van Lindert and Doutje Lettinga (eds), *The Future of Human Rights in an Urban World: Exploring Opportunities, Threats and Challenges* (Amnesty International Netherlands 2014), pg. 14.

[38] Ibid.

[39] Heidi Hobbs, *City Hall Goes Abroad: The Foreign Policy of Local Politics* (Sage Publishing 1994), pg. 5.

[40] I borrow the term 'bringing international law home' from Harold H. Koh, 'The 1998 Frankel Lecture: Bringing International Law Home' (1998) 35 *Houston Law Review* 623.

[41] UNFCCC, 'Status of Ratification of the Kyoto Protocol', online: http://unfccc.int/kyoto_protocol/status_of_ratification/items/2613.php (accessed on 1 May 2016).

[42] Greg Kahn, 'The Fate of the Kyoto Protocol under the Bush Administration' (2003) 21 *Berkeley Journal of International Law* 548, pgs. 551–559; Julian Borger, 'Bush Kills Global Warming Treaty' *Guardian* (29 March 2001), online: www.theguardian.com/environment/2001/mar/29/globalwarming.usnews (accessed on 1 May 2016).

Congress to pass legislation that would establish a national emissions trading system.[43] As of 14 December 2015, 1,060 cities across all American states have signed the agreement, thereby internalizing an international law instrument, the Kyoto Protocol, within the United States despite non-ratification by the federal government.

On 1 June 2017, United States President Donald Trump announced that his country 'will cease all implementation of the non-binding Paris Accord' and that he is 'willing to . . . either negotiate [the country's] way back into Paris, under terms that are fair to the United States and its workers, or to negotiate a new deal that protects [the United States] and its taxpayers'.[44] Within a few days of President Trump's announcement, more than a thousand governors, mayors, and business leaders declared that regardless of what the federal government might do, American cities, states, and businesses will strive to fulfil the United States' commitments under the Paris Agreement.[45] Amongst American cities, New York City positioned itself as the main bulwark against Trump's 'climate stupidity'.[46] Bill de Blasio, New York City's mayor, signed an executive order the day after Trump's decision, announcing that his city 'will adopt the principles and goals of the Paris Agreement'.[47] The mayor also directed city agencies to work with their national counterparts, 'global climate network partners and other leading cities to develop further [GHG] reduction plans and actions that are consistent with the principles and goals of the Paris Agreement . . . to meet [America's] 2016 commitment under the Paris Agreement'.[48] The mayor of Pittsburgh, William Peduto, also issued an executive order declaring that his city 'endorses and remains fully committed to the principles of the Paris Agreement'.[49] Mayor Peduto further declared that

[43] US Conference of Mayors Climate Protection Agreement, online: www.usmayors.org/climateprotection/agreement.htm (accessed on 1 May 2016).

[44] The White House, Office of the Press Secretary, 'Statement by President Trump on the Paris Climate Accord', 1 June 2017, online: www.whitehouse.gov/the-press-office/2017/06/01/statement-president-trump-paris-climate-accord (accessed on 1 August 2017).

[45] California Governor Jerry Brown and UN Secretary-General's Special Envoy for Cities and Climate Change Michael Bloomberg launched an initiative, 'America's Pledge on Climate Change', to compile and quantify the actions of cities, states, and businesses in the United States to reduce their GHG emissions consistent with the goals of the Paris Agreement; 'Jerry Brown and Michael Bloomberg Launch "America's Pledge" in Support of Paris Agreement', online: http://newsroom.unfccc.int/unfccc-newsroom/jerry-brown-and-michael-bloomberg-launch-americas-pledge-in-support-of-paris/ (accessed on 1 August 2017). For discussion, see Robinson Meyer, 'America's Pledge: Can States and Cities Really Address Climate Change?' *Atlantic* (2 June 2017).

[46] *New York Times* Editorial Board, 'States and Cities Compensate for Mr. Trump's Climate Stupidity' *New York Times* (7 June 2017).

[47] The City of New York, Office of the Mayor, 'Executive Order No. 26: Climate Action Executive Order', 2 June 2017, online: www1.nyc.gov/assets/home/downloads/pdf/executive-orders/2017/eo_26.pdf (accessed on 1 August 2017).

[48] Ibid.

[49] City of Pittsburgh, Office of the Mayor, 'Executive Order No. 2017–08: Reinforcing Pittsburgh's Commitment to the Global Partnership on Climate Change', 2 June 2017, online: http://apps.pittsburghpa.gov/mayorpeduto/Climate_exec_order_06.02.17_(1).pdf (accessed on 1 August 2017).

Pittsburgh's chief resilience officer is empowered to undertake additional actions to meet the 1.5-degree Celsius target as set out in the Paris Agreement.[50]

The third example takes us away from the international climate change agreements. In 1981, women's rights activists reached a remarkable milestone when the Convention on the Elimination of All Forms of Discrimination Against Women (CEDAW) entered into force.[51] A summary of the treaty's objectives can be found in Article 3, which requires signatory states to 'take in all fields, in particular in the political, social, economic and cultural fields, all appropriate measures, including legislation, to ensure the full development and advancement of women, for the purpose of guaranteeing them the exercise and enjoyment of human rights and fundamental freedoms on a basis of equality with men'.[52] US President Carter signed CEDAW in 1980, but subsequent administrations either have failed or not tried to secure ratification of the treaty by the Senate.[53] Many opponents of CEDAW ratification are concerned that the treaty threatens American norms and values, particularly those concerning the structure of the family,[54] abortion,[55] and family planning.[56] Often using provocative language, CEDAW's critics have highlighted the treaty's challenge to a conception of women as obliged, first and foremost, to their households.[57]

As with the case of climate change, cities have not acquiesced. As Resnik and colleagues put it, 'a few went beyond expressive statements and aimed to turn "transnational" law into "local" law'.[58] In 1998, San Francisco passed a local ordinance to implement CEDAW.[59] The city committed itself to the CEDAW technique of lawmaking through self-reflective enquiry about the effects of equality norms across all domains. Specifically, it strives to '[r]eview federal, state and local laws and public policies to identify systematic and structural discrimination against women and girls' and ultimately ensure that '[e]verything that happens to San Francisco women and girls will be interpreted and acted upon using the CEDAW conceptual framework, analysis and language'.[60] San Francisco has inspired other cities – Portland (Oregon) and Berkeley (California) – to pass similar legislation. In 2013, an NGO launched the 'Cities for CEDAW' campaign which aims 'to "make

[50] Ibid. [51] 1249 UNTS 13. [52] Ibid.

[53] Luisa Blanchfield, *The U.N. Convention on the Elimination of All Forms of Discrimination Against Women (CEDAW): Issues in the U.S. Ratification Debate* (Congressional Research Service, 28 June 2011), online: www.fas.org/sgp/crs/row/R40750.pdf, pg. 1.

[54] Ibid., pg. 13. [55] Ibid., pg. 15. [56] Ibid., pg. 158.

[57] Judith Resnik, Joshua Civin, and Joseph Frueh, 'Ratifying Kyoto at the Local Level: Sovereigntism, Federalism, and Translocal Organizations of Government Actors (TOGAs)' (2008) 50 *Arizona Law Review* 709, pg. 724.

[58] Ibid.

[59] For discussion of San Francisco's ordinance, see Stacy Laira Lozner, 'Diffusion of Local Regulatory Innovations: The San Francisco CEDAW Ordinance and the New York City Human Rights Initiative' (2004) 104 *Columbia Law Review* 768.

[60] City and County of San Francisco, Department on the Status of Women, 'CEDAW Action Plan', online: http://sfgov.org/dosw/cedaw-action-plan (accessed on 1 May 2016).

the global local" by harnessing the power of cities and promoting the adoption of CEDAW as a municipal ordinance in cities ... in order to create a framework for improving the status of women and girls'.[61]

Finally, there is a growing transnational movement of 'Human Rights Cities' as civil society, cities, and international organizations try to realize the achievement of human rights at the local level. Through their comments, declarations, and statements, UN supervisory bodies and regional institutions have explicitly recognized the role of local authorities in giving effect to human rights.[62] Human rights cities can be defined as 'cities that explicitly refer to human rights norms in their activities, statements or policy'.[63] Van den Berg and Oomen have shown that there are many approaches that cities can take to implement international human rights. Some cities use the methodology developed by the NGO known as The People's Movement for Human Rights Leaning (PDHRE).[64] Key aspects of PDHRE's methodology include human rights education, the formation of a steering group in the city, the development and implementation of action plans, and evaluation of activities to foster awareness and promote realization of human rights by various stakeholders in society. Human rights cities that have put this methodology into practice include Rosario (Argentina), Graz (Austria), and Nagpur (India).[65] Some cities, like Utrecht in the Netherlands, base their policies on the general idea of human rights and seek to incorporate all international and regional human rights instruments in their local policies.[66] Van den Berg and Oomen have argued that '[i]n referring to international human rights as a basis for their policies, cities can also demarcate their autonomy, and become part of a powerful network of global actors instead of being subservient to the nation states. This process of "glocalization" also entails a new type of citizenship that straddles the local and the global.'[67]

3.2.2 *City Diplomacy*

Diplomacy is predominantly perceived to be within the realm of interstate relations. The *Oxford Encyclopedia of the Modern World*, for example, defines diplomacy as 'the formalized system of procedures or the process by which sovereign states, usually through ambassadors or other representatives, conduct their official

[61] Cities for CEDAW, 'Implementing CEDAW as a Local Ordinance', online: http://citiesforcedaw.org/wp-content/uploads/2015/01/Fact-Sheet-on-Implementing-CEDAW-as-a-Local-Ordinance.pdf (accessed on 1 May 2016).

[62] See discussion in Antoine Meyer, 'Local Governments and Human Rights Implementation: Taking Stock and a Closer Strategic Look' (2009) 6 *Pace Diritti Umani* 7, pgs. 11–13.

[63] Van den Berg and Oomen, pg. 13.

[64] PDHRE, 'Human Rights Cities – A Practical Way to Learn and Chart the Future of Humanity', online: www.pdhre.org/projects/hrcommun.html (accessed on 1 May 2016).

[65] Ibid.

[66] Gemeente Utrecht, *Mensenrechten in Utrecht: hoe geeft Utrecht invulling aan internationale mensenrechtenverdragen? Een stedelijke zoektocht naar sociale rechtvaardigheid* (2011).

[67] Van den Berg and Oomen, pg. 15.

relations'.[68] Within diplomatic studies, a niche literature on the concept of 'parallel diplomacy' or 'paradiplomacy' considers the possibility of a number of external relations 'tracks' running across countries.[69] According to the paradiplomacy concept, these tracks do not only involve traditional diplomatic actors such as the State Department or the Ministry of Foreign Affairs. Some tracks will constitute the external relations of sub-national actors. The literature on paradiplomacy tends to be focused on regions, provinces, and states within federalist systems.[70] The diplomatic role of cities, let alone that of global cities, has received little consideration in the literature to date.

Since the mid 2000s, however, there has been increasing discussion of inter-city diplomacy. In 2006, the Clingendael (Netherlands Institute of International Relations) launched a pilot project to provide an overview of the burgeoning landscape of diplomatic developments taking place at the level of cities.[71] This project defined *city diplomacy* as 'the institutions and processes by which cities engage in relations with actors on an international political stage with the aim of representing themselves and their interests to one another'.[72] From this point of view, diplomacy is not confined to ambassadorial and political advocacy activities but also encompasses trade facilitation and cultural exchange. It should be noted that United Cities Local Governments' (UCLG) Committee on City Diplomacy, Peace Building and Human Rights, established in 2005, defines 'city diplomacy as the tool of local governments and their associations for promoting social cohesion, conflict prevention, conflict resolution and post-conflict reconstruction with the aim of creating a stable environment, in which the citizens can live together in peace, democracy and prosperity'.[73] This definition emphasizes the role of cities in conflict resolution and

[68] Peter Stearns (ed), *Oxford Encyclopedia of the Modern World* (Oxford University Press 2008).

[69] See, for example, Alexander S. Kuznetsov, *Theory and Practice of Paradiplomacy: Subnational Governments in International Affairs* (Routledge 2014); Ivo D. Duchacek, Daniel Latouche, and Garth Stevenson (eds), *Perforated Sovereignties and International Relations: Trans-Sovereign Contacts and Subnational Governments* (Greenwood Press 1988); Carlos R. S. Milani and Maria Clotilde Meirelles Ribeiro, 'International Relations and the Paradiplomacy of Brazilian Cities: Crafting the Concept of Local International Management' (2011) 8 *Brazilian Administration Review* 21.

[70] See, for example, Hans J. Michelmann and Panayotis Soldatos (eds), *Federalism and International Relations: The Role of Subnational Units* (Oxford University Press 1990); Hubert Rioux Quimet, 'From Sub-State Nationalism to Subnational Competition States: The Development and Institutionalization of Commercial Paradiplomacy in Scotland and Quebec' (2015) 25 *Regional & Federal Studies* 109; Michael Keating, 'Regions and International Affairs: Motives, Opportunities and Strategies' (1999) 9 *Regional and Federal Studies* 1.

[71] Rogier van der Pluijm and Jan Melissen, *City Diplomacy: The Expanding Role of Cities in International Politics* (Clingendael Institute, 1 April 2007), online: www.clingendael.nl/publication/city-diplomacy-expanding-role-cities-international-politics (accessed on 20 December 2016), pg. 11. Also see the 'Shanghai Consensus on the Role of Cities in International Relations', online: www.clingendael.nl/news/shanghai-consensus-cities-international-relations (accessed on 20 December 2016).

[72] Ibid., pg. 5.

[73] The Hague Agenda on City Diplomacy, 13 June 2008, online: www.peaceprize.uclg.org/documents-of-interest.html (accessed on 3 April 2015).

peace building. The committee merged with the Decentralized Cooperation Committee in 2010,[74] and the new entity, known as the Committee on Development Cooperation and City Diplomacy, has a broader mandate of promoting development cooperation, achieving the Millennium Development Goals (which were replaced by the Sustainable Development Goals in 2015), and advising UCLG on the role of cities in preventing violent conflicts.[75]

In some regions such as East Asia and Europe, there has been a rise in inter-city diplomatic activities to achieve a range of objectives including the easing of long-standing political tensions and trade promotion.[76] For example, in April 2014, the media reported that Tokyo's governor would visit Beijing at the invitation of its mayor to help 'heal Japan and China's bruised relations'.[77] On the agenda was Tokyo sharing its experience of tackling air pollution (which is a serious problem in the Chinese capital) and, in turn, learning more about organizing the Olympics Games from the Beijing mayoralty. Underlying this agenda of sharing best practices and inter-city learning is the quest by Beijing and Tokyo to 'help improve Japan-China relations by building on these exchanges'.[78] A second example is that of the Taipei city government. A section on the city government's website is devoted to documenting the 'city diplomacy' that Taipei engages in to enhance 'the attitude and vision of the global city that Taipei has become'.[79] A notable initiative is the Taipei-Shanghai City Forum, which serves as a platform for advancing city-to-city ties amidst continuing deep divisions in relations between Taiwan and the People's Republic of China.[80] The forum, which has taken place annually since 2010, focuses on cultural and economic issues, and rarely touches on political matters.[81]

A third contemporary example of city diplomacy concerns how cities around the world reacted to the passage of anti-gay legislation in St Petersburg, Russia. On 30 January 2013, the *Huffington Post* and other news agencies reported that Venice's City Council had voted to suspend its relationship with St Petersburg after the city

[74] UCLG, 'Committees and Working Groups: Development Cooperation and City Diplomacy', online: www.uclg.org/en/organisation/structure/committees-working-groups/development-cooperation-and-city-diplomacy (accessed on 3 December 2016).

[75] Ibid.

[76] Outside Europe and East Asia, inter-city diplomacy has also been recognized for its role in building peace in the Middle East; see, for example, European Commission, *Palestinian and Israeli Mayors Praise Role of City-to-City Diplomacy in Middle East Peace Process* (1 June 2007), online: http://europa.eu/rapid/press-release_COR-07–73_en.htm (accessed on 3 April 2016).

[77] ' Tokyo Governor to Make "City Diplomacy" Visit to Beijing' *Straits Times* (15 April 2014).

[78] Ibid.

[79] Taipei City Government, 'City Diplomacy', online: http://english.sec.gov.taipei/np.asp?ctNode=84311&mp=101002 (accessed on 1 May 2016). This statement is indicative of how some cities view external relations as a means of demonstrating its global city status.

[80] Russell Flannery, 'Taipei Mayor Ready to 'Dance with Wolves' in Shanghai Next Week' *Forbes* (14 August 2015); Pichi Chuang, 'Taipei Mayor Won't Publicly Back "One China", Shanghai Forum at Risk' *Reuters* (30 July 2015).

[81] 'Taipei Mayor Ko Wen-je to Visit Shanghai after Voicing "Respect" for "One China" Consensus' *South China Morning Post* (5 August 2015).

passed anti-gay legislation that violates global norms concerning lesbian, gay, bisexual, and transgender (LGBT) rights.[82] The law criminalizes 'public action aimed at propagandising sodomy, lesbianism, bisexualism, and transgenderism among minors'.[83] Those charged with breaking the law face fines amounting from 5,000 to 500,000 Russian rubles (approximately 68 to 6,781 euros).[84]According to news reports, the Venice City Council voted to suspend its 2006 cultural cooperation agreement with St Petersburg as long as the anti-gay legislation remained in force.[85] As Nijman has pointed out, instead of the Italian state responding to the human rights violation by another member state of the Council of Europe, this is a case of an Italian city upholding global human rights norms.[86] Venice's actions also prompted citizens in other cities such as Melbourne (Australia), Los Angeles (US), and Manchester (UK) to urge their city governments to suspend their twin-city status with St Petersburg.[87] The response of Melbourne's city government is also noteworthy from the perspective of city diplomacy. Although the city council decided against suspending Melbourne's relationship with St Petersburg, it justified its decision on the basis that maintaining the relationship with St Petersburg would allow Melbourne to more actively advocate for the revocation of the anti-gay law.[88]

City diplomacy often takes the form of developing trade links between cities. For example, the cities of Istanbul (Turkey) and Melbourne (Australia) have expressed their commitment to business-oriented city diplomacy. The Greater Istanbul municipality's Office for External Relations cooperates with the Istanbul Chamber of Commerce to organize business delegation trips abroad.[89] It also provides advisory services to support foreign investors in Istanbul.[90] Melbourne has an 'International Engagement Framework', endorsed by the city's council in April 2010, which

[82] Glennisha Morgan, 'Venice to Cut Ties with St. Petersburg over Anti-Gay Law' *Huffington Post* (30 January 2013), online: www.huffingtonpost.com/2013/01/30/venice-st-petersburg-anti-gaw-propaganda-law_n_2576044.html (accessed on 1 May 2016).

[83] Miriam Elder, 'St Petersburg Bans "Homosexual Propaganda"' *Guardian* (12 March 2012), online: www.theguardian.com/world/2012/mar/12/st-petersburg-bans-homosexual-propaganda (accessed on 1 May 2016).

[84] Ibid [85] Morgan.

[86] Janne E. Nijman, 'Renaissance of the City as Global Actor: The Role of Foreign Policy and International Law Practices in the Construction of Cities as Global Actors' in Gunther Hellmann, Andreas Fahrmeir, and Milos Vec (eds), *The Transformation of Foreign Policy: Drawing and Managing Boundaries from Antiquity to the Present* (Oxford University Press 2016), pg. 1.

[87] Claire Bigg, 'Sister Cities Ramp Up Russia Boycott Over Antigay Law' *Radio Free Europe* (19 July 2013), online: www.rferl.org/content/russia-sister-cities-gay-law/25051513.html (accessed on 1 May 2016); 'Manchester to Use St Petersburg Link to "Pressure Russia"' *BBC News* (2 August 2013), online: www.bbc.com/news/uk-england-manchester-23536652 (accessed on 1 May 2016).

[88] Beau Donelly, 'Melbourne Won't Sever Ties with Sister City over Anti-Gay Laws' *Weekly Review* (2 July 2013), online: www.theweeklyreview.com.au/uncategorized/1611790-melbourne-wont-sever-ties-with-sister-city-over-anti-gay-laws/; Jason Dowling, 'Lord Mayor Tackles Russia over Gay Rights' *Age* (7 March 2013), online: www.theage.com.au/victoria/lord-mayor-tackles-russia-over-gay-rights-20130306-2flqg.html (both accessed on 1 May 2016).

[89] 'City Diplomacy and Istanbul' *Turkish Review* (1 August 2013). [90] Ibid.

indicates the priority areas for the city's international relations and the amount of resources that ought to be devoted to each priority area.[91] It is noteworthy that the framework sets out that the city ought to devote 50 per cent of its city diplomatic resources to 'building prosperity', which is defined as '[increasing] export and inward investment in Melbourne's key industry sectors, including education, tourism and services sector', 20 per cent towards increasing inter-city learning, 20 per cent to supporting community and cultural programmes, and 10 per cent towards participation in international governance.[92] Melbourne is also the only Australian city to operate an overseas business representative office in Tianjin, China, to strengthen trade links between the two cities.[93]

A final example is that of the European Network of Local Authorities for Peace in the Middle East. This network of city governments organizes many lobbying activities such as election monitoring and aims to keep the Middle East peace process on the international agenda.[94] Proposed by the Association of Palestinian Local Authorities and the Union of Local Authorities in Israel, the network also serves as a forum for Israeli–Palestinian municipal dialogue, with contributions from European cities (including Rome, Barcelona, and Cologne).[95] Cooperation is often based on trilateral development projects in the areas of culture, youth, the environment, and municipal management.[96]

As this discussion indicates, economic activity drives quite a few instances of modern inter-city diplomacy, a circumstance that is not unexpected. In Europe, Venetian and Byzantine commercial diplomacy dates back to the early Middle Ages.[97] From the twelfth century onwards, commerce became a crucial topic of diplomatic negotiations.[98] Today, diplomacy with an eye towards trade facilitation and establishing economic contact between cities can be explained by the decreasing amount of funding allocated by central governments to their cities and the consequent need for cities to generate independent sources of income. Whereas decentralization and the devolution of political authority from higher levels of government to cities lead to municipal governments having more autonomy to

91 City of Melbourne, *Our International Strategy – Melbourne: Doing Business Globally* (2014), online: www.melbourne.vic.gov.au/business/doing-business-globally/Pages/melbourne-doing-business-globally.aspx (accessed on 1 May 2016).

92 Ibid., pg. 1.

93 Melbourne Office Tianjin, online: www.melbourne.vic.gov.au/business/doing-business-globally/international-programs/Pages/melbourne-office-tianjin.aspx (accessed on 1 May 2016).

94 Alexandra Sizoo and Arne Musch, 'City Diplomacy: The Role of Local Governments in Conflict Prevention, Peace-Building and Post-Conflict Reconstruction' in Arne Musch et al. (eds), *City Diplomacy: The Role of Local Governments in Conflict Prevention, Peace-Building, Post-Conflict Reconstruction* (VNG International, The Hague, 2008), pg. 17.

95 Chris van Hemert, 'A Case Study in City Diplomacy: The Municipal Alliance for Peace in the Middle East' in Arne Musch et al. (eds), *City Diplomacy: The Role of Local Governments in Conflict Prevention, Peace-Building, Post-Conflict Reconstruction* (VNG International, The Hague, 2008), pg. 166.

96 Ibid. 97 Stearns. 98 Ibid.

establish trade and diplomatic links with their counterparts, decentralization also often involves the higher levels of government delegating more responsibilities to cities without increasing the cities' budgets.[99] This can lead to failed decentralization in some instances.[100] In others, cities manage to avoid local bankruptcy and fill their coffers by successfully attracting foreign investment and creating markets for the businesses that reside in their cities.[101] Political activities represent a new direction for cities as they venture beyond their jurisdictional boundaries to make statements on more far-reaching issues such as human rights and nuclear disarmament.

3.2.3 *Cities Developing Independent Strategies to Manage Global Threats*

Cities today face many global risks, such as infectious diseases and terrorism.[102] The discussion in this section focuses on counterterrorism efforts because it is in this area that cities have developed notably sophisticated responses. According to Hank Savitch, nearly three-quarters of incidents labelled as terror attacks worldwide, and four out of every five of their subsequent casualties, occur in cities.[103] Savitch argues that the complexity of the urban landscape makes it ideal for hiding terrorist plots.[104] Densely populated cities also facilitate extensive loss of life and damage to property, generating strong symbolic meaning, fear, and anxiety. Counter-insurgency expert David Kilcullen argues that cities are the target: 'The goal is to shut [cities] down for as long as possible, separate people from one another, break down communities, and push them into mental fortresses.'[105] The breakdown of trust in communities facilitates further exploitation.

[99] The goal of decentralization is to disperse authority and responsibility to lower levels of decision-making. In theory, the two main benefits of decentralization are allocative efficiency and improved governance. For discussion, see Mark Turner, David Hulme, and Willy McCourt, *Governance, Management and Development: Making the State Work* (Palgrave Macmillan 2015), chapter 8. For an excellent discussion of the Indian experience with decentralization, see Sharmila L. Murthy and Maya J. Mahin, 'Constitutional Impediments to Decentralization in the World's Largest Federal Country' (2015) 26 *Duke Journal of Comparative and International Law* 79.

[100] Ibid.

[101] Earl H. Fry, 'State and Local Governments in the International Arena' (1990) 509 *Annals of the American Academy of Political and Social Science* 118, pgs. 122–123.

[102] On the risks of epidemics and infectious diseases that a global city faces, see Lance Saker et al., *Globalization and Infectious Diseases: A Review of the Linkages* (UNDP/World Bank/WHO Special Programme for Research and Training in Tropical Diseases 2004); Harris Ali and Roger Keil, 'Global Cities and the Spread of Infectious Disease: The Case of Severe Acute Respiratory Syndrome (SARS) in Toronto, Canada' (2006) 43 *Urban Studies* 491; Roger Keil and Harris Ali, 'Governing the Sick City: Urban Governance in the Age of Emerging Infectious Disease' (2007) 39 *Antipode* 846.

[103] Hank V. Savitch, *Cities in a Time of Terror: Space, Territory, and Local Resilience* (Routledge 2008), pgs. 3–7.

[104] Ibid.

[105] Robert Muggah, 'Is Urban Terrorism the New Normal? Probably' *World Economic Forum, Davos 2016* (17 January 2016), online: www.weforum.org/agenda/2016/01/is-urban-terrorism-is-the-new-normal-probably/ (accessed on 5 May 2016).

Although terrorism has traditionally been perceived to be a matter of national security, increasingly global cities are acting autonomously as they develop and institutionalize their own local and transnational counterterrorism strategies.[106] New York City, the city at the heart of the September 11 attacks, offers a prime example of how a global city has forged ahead with internationalizing its municipal policing strategy and forming transnational partnerships based on the belief that cities 'must serve as the frontline of homeland security' and cannot rely on the federal government to provide adequate protection.[107]

The New York City Police Department (NYPD) is the largest municipal police department in the United States.[108] With about 36,000 officers and 15,000 support staff, the department is twice the size of the Federal Bureau of Investigation (FBI), which is the federal counterterrorism agency.[109] Prior to the September 11 attacks, the NYPD focused almost solely on crime reduction, but after the attacks, it was fundamentally reorganized and counterterrorism was made one of the NYPD's key priorities.[110] An entirely new Counterterrorism Bureau was created in 2002, and the existing intelligence division was revamped. The Counterterrorism Bureau employs 250 full-time officers, of which about half are part of the New York Joint Terrorism Task Force, along with the FBI and other federal law enforcement agencies.[111] These officers have security clearance that gives them access to national and international sources of intelligence and to investigations conducted overseas.[112] Through the New York Joint Terrorism Task Force, NYPD officers have been able to interrogate terrorist suspects in Afghanistan and Pakistan and to conduct interviews at Guantanamo Bay.[113] The Counterterrorism Bureau provides some level of counterterrorism training for all NYPD officers, such as recognizing suspicious behaviour and the use of gear that protects against biological, chemical, and radioactive weapons.[114] The Counterterrorism Bureau is most visible when it carries out the

[106] Kristin Ljungkvist, *Global City 2.0: From Strategic Site to Global Actor* (Routledge 2016), pgs. 77–78. A few hours after the events of September 11, New York City Mayor Rudolph Giuliani and New York Governor George Pataki jointly gave a press conference. Mayor Giuliani indicated that it was not up to him or the city of New York but to the US president to take charge of the situation and respond. As Ljungkvist notes, this is not surprising because, up till then, counterterrorism was deemed to be a federal concern. This perception would eventually shift, as will be noted later in this chapter.

[107] For an articulation of this view, see, for example, Eben Kaplan, 'New York Spurs Counterterrorism Efforts' *Council on Foreign Relations* (28 December 2006), online: www.cfr.org/world/new-york-spurs-counterterrorism-efforts/p12312; the mission statement of the New York Police Department Counterterrorism Bureau states: 'Built upon the realization that the City could not rely solely on the federal government for its defense, the Counterterrorism Bureau was created', online: www.nyc.gov/html/nypd/html/administration/counterterrorism_units.shtml (both websites accessed on 3 May 2016).

[108] Office of Justice Programs, Bureau of Justice Statistics, latest data available: 2008; see Brian Reaves, *Census of State and Local Law Enforcement Agencies, 2008* (US Department of Justice, July 2011), pg. 4, online: www.bjs.gov/index.cfm?ty=dcdetail&iid=249 (accessed on 3 May 2016).

[109] Brian Nussbaum, 'Protecting Global Cities: New York, London and the Internationalization of Municipal Policing for Counter Terrorism' (2007) 8 *Global Crime* 213, pg. 218.

[110] Ljungkvist, pg. 78. [111] Ibid. [112] Ljungkvist, pg. 75. [113] Ibid. [114] Nussbaum, pg. 219.

massive deployment of heavily armed, paramilitary-style units at high-profile locations around the city. According to Nussbaum, NYPD senior officials stress that these deployments are not random, but driven by intelligence and have proven to be effective.[115]

In addition to the establishment of the Counterterrorism Bureau, under the leadership of Mayor Bloomberg and Police Commissioner Ray Kelly, significant resources and expertise were put into revamping the NYPD's intelligence division into one 'that rivals the security services of many small countries'.[116] The division is staffed by approximately 800 people. About half of the division focuses specifically on terrorism, while the other half focuses on criminal activity such as drug trafficking and gang violence.[117] The counterterrorism division has an International Liaison Program whereby NYPD detectives are sent to live in Europe, the Middle East, and Southeast Asia, where they serve as the NYPD's liaisons to other countries' law enforcement and intelligence communities.[118] These officers are not armed and do not become directly involved in investigations and enforcement actions in their host countries; their primary objective is intelligence gathering.[119] Since the inception of the International Liaison Program, there has been a senior NYPD officer at the scene of most terrorist attacks in the world – such as the ones in Istanbul, Madrid, and Jakarta – to assess how the attack might be relevant to New York City.[120] It is noteworthy that the NYPD's overseas expansion has been driven mainly by the desire to work around and bypass the FBI. The need for New York City to develop its local counterterrorism capabilities has been rationalized on the basis that the federal government has proven incapable of protecting the city and that the city's government has to step in and take on the responsibility of providing security for its citizens.[121]

It can be argued that, regardless of whether terrorism is perceived to be a local or international threat, most cities do not have the resources or the autonomy to address it at the global level. The US$200 million that New York City spends annually on counterterrorism alone vastly exceeds the budgets of most cities.[122] In this respect, New York City is an outlier. Although this may well be the case, it does not detract from the main point advanced here: i.e. that there are a handful of global cities, of which New York and London[123] are examples, that have developed sophisticated local and transnational counterterrorism capabilities independent of their national governments. Furthermore, a number of cities such as Chicago and Los Angeles,

115 Nussbaum, pg. 220. 116 Ibid. 117 Ljungkvist, pg. 72.

118 This programme is funded by the New York City Police Foundation, a non-profit group that is backed by private multinational corporations such as JP Morgan, Goldman Sachs, and Barclays; www.nycpolicefoundation.org/programs/international-liaison-program/ (accessed on 5 December 2016).

119 New York City Police Foundation, 'International Liaison Program', online: www.nycpolicefoundation.org/programs/international-liaison-program/ (accessed on 5 December 2016).

120 Ibid. 121 Ljungkvist, pg. 107. 122 Kaplan.

123 SeeJon Coaffee, *Terrorism, Risk and the Global City: Towards Urban Resilience* (Ashgate Publishing 2009), for a detailed study of London's responses to terrorism in the 1990s.

inspired by New York City, are increasingly developing their autonomous counter-terrorism capabilities.[124]

3.2.4 Global Organizations and Global Aims

In *Expanding Governmental Diversity in Global Governance: Parliamentarians of States and Local Governments*, Alger provides an overview of international organizations of cities and local governments to 'challenge readers to realize that global governance is ever more complicated and does not only involve the UN system and organizations involved in the UN system'.[125] He categorizes 'global organizations of local authorities' according to five foci – general purpose, larger cities, environmental, peace, and language – a classification which I follow in this section.[126]

General Purpose: An example is UCLG, which aims to 'increase the role and influence of local governments in global governance' and facilitates programmes and partnerships that build the capacities of local governments.[127] Since its creation in 2004, UCLG has advocated for a formal advisory role for local government within the UN system. One of the objectives adopted at UCLG's Founding Congress is 'renewing and deepening our partnership with United Nations and the global community, and building an effective and formal role for local government as a pillar of the international system'.[128] Indeed, it can be argued that UCLG won the highest accolade of recognition as the unified voice of cities worldwide when the 2004 *Report of the Panel of Eminent Persons on United Nations-Civil Society Relations* (the Cardoso Report) made the specific recommendation that 'the United Nations should regard United Cities and Local Governments as an advisory body on governance matters'.[129]

Larger cities: Created in 1985, Metropolis (World Association of the Major Metropolises) brings together cities and metropolitan regions with more than a million inhabitants.[130] Its mission is to serve 'as an international forum for exploring issues and concerns common to all big cities and metropolitan regions'.[131] Metropolis also manages the metropolitan section of UCLG. As of October 2014,

124 Kaplan.

125 Chadwick F. Alger, 'Expanding Governmental Diversity in Global Governance: Parliamentarians of States and Local Governments' (2010) 16 *Global Governance* 59, pg. 60.

126 Ibid., pg. 63.

127 UCLG, 'About Us', online: www.uclg.org/en/organisation/about (accessed on 18 April 2016).

128 UCLG, 'International Agenda', online: www.uclg.org/en/issues/united-nations-advocacy (accessed on 18 April 2016).

129 *We the Peoples: Civil Society, the United Nations and Global Governance* (Panel of Eminent Persons on UN-Civil Society Relations, UN General Assembly, Fifty-eighth session, Agenda item 59: Strengthening of the United Nations System, Doc A/58/817, 11 June 2004), pg. 20. For discussion about the Cardoso Report, see Peter Willetts, 'The Cardoso Report on the UN and Civil Society: Functionalism, Global Corporatism, or Global Democracy?' (2006) 12 *Global Governance: A Review of Multilateralism and International Organizations* 305

130 Metropolis, 'Mission', www.metropolis.org/mission (accessed on 18 April 2016).

131 Ibid.

it has 139 active members across all geographical regions, including Dubai, Jakarta, Madrid, Mexico City, Moscow, and Toronto.[132]

Environmental: ICLEI is one of the world's largest organizations of local governments. Its membership in 2012 included 12 megacities, 100 super-cities and urban regions, and 450 large cities, as well as another 450 small and mid-sized cities in 84 countries.[133] Formed in 1990 by the International Union of Local Authorities and the United Nations Environment Programme (UNEP) to represent the environmental concerns of local government internationally, the network was formerly known simply as the International Council for Local Environmental Initiatives. However, in order to highlight the network's approach of 'look[ing] beyond mere environmental aspects and embrac[ing] wider sustainability issues', the ICLEI council renamed the association ICLEI – Local Governments for Sustainability.[134] Amongst other things, ICLEI works to promote biodiversity conservation and resource efficiency at the local level.[135] Its Cities Biodiversity Center works closely with the Convention on Biological Diversity (CBD) secretariat to organize capacity-building events for cities and side events parallel to COP meetings. On resource efficiency, ICLEI focuses on helping cities better manage their natural resources such as water and soil. ICLEI represents its members in major sustainability forums such as UN-Water, the International Water Association, and UN Habitat.[136]

Peace: Mayors for Peace was established in 1982 by the then-mayor of Hiroshima, who proposed a new Program to Promote the Solidarity of Cities toward the Total Abolition of Nuclear Weapons at the second UN Special Session on Disarmament held at the UN Headquarters in New York.[137] The organization aims to promote the abolition of nuclear weapons; its 2020 Vision Campaign has the ambitious goal of abolishing nuclear weapons by the year 2020.[138] A total of 6,649 cities are members of Mayors for Peace as of 1 April 2015.[139] The role of cities in conflict prevention, peace

[132] Metropolis, 'List of Active Members', online: www.metropolis.org/sites/default/files/pdf/list_of_active_members.pdf (accessed on 18 April 2016).

[133] ICLEI, *ICLEI – Local Governments for Sustainability Corporate Report 2011/12* (2012), pg. 7, online: www.iclei.org (accessed on 3 December 2016). Megacities are cities with a population of at least 10 million people. In ICLEI terminology, a super city has a population of 1 to 10 million people, a large city has a 100,000 to 1 million people, and a city or town has up to 100,000 people.

[134] ICLEI, 'Who Is ICLEI', online: www.iclei.org/iclei-global/who-is-iclei.html (accessed on 3 December 2016).

[135] Ibid. [136] Ibid.

[137] Mayors for Peace, 'About Us', online: www.mayorsforpeace.org/english/outlines/index.html (accessed on 3 December 2016).

[138] 2020 Vision Campaign, online: www.2020visioncampaign.org/en/about-us.html (accessed on 3 December 2016).

[139] Mayors for Peace, 'Member Cities', online: www.mayorsforpeace.org/english/membercity/map.html (accessed on 3 December 2016). Van der Pluijm and Melissen point out that, despite the widespread support that Mayors for Peace enjoys, 'the mayors have not been able to stop the process of nuclear proliferation in various countries around the world'; pg. 21.

building, and post-conflict reconstruction has already been discussed earlier within the context of city diplomacy.

Language: Founded in 1979, the Association Internationales des Maires Francophones (AIMF), or the International Association of Francophone Mayors, brings together mayors and officials from cities where French is the official language or is widely used.[140] Its members include Siem Reap and Phnom Penh (Cambodia), Edea and Garoua (Cameroon), Brazzaville (Congo), Paris and Lille (France), Montreal (Canada), Geneva and Lausanne (Switzerland), and Hanoi (Vietnam).[141] AIMF's programmes aim to build capacity amongst local officials, assist cities in raising funds to provide essential public services, provide policy training and tools to improve local finances, and promote heritage conservation in cities.[142]

These examples are hardly exhaustive but provide a flavour of the wide range of organizations that cities have created to pursue common aims. These organizations are global in reach and often seek to tackle global challenges such as environmental protection and gender equality; they are examples of how cities are organizing themselves to represent their interests or pursue shared objectives.

3.3 THE GLOBAL URBAN AGENDA

Two international organizations play a key role in setting and implementing the global agenda on urbanization: the World Bank and UN-Habitat. By exploring the ideas underpinning their urban policies, this section argues that the World Bank and UN-Habitat have created and are sustaining an ideological narrative that supports the rise of cities in international affairs. The values and interests advanced by the World Bank and UN-Habitat in their ideological narrative coincide with those of cities – that is, that there ought to be greater decentralization of political authority from central governments to cities so that cities can have greater autonomy and ultimately a role to play in global governance processes. Through its lending policies and technical assistance programmes, the World Bank is able to put its ideas into practice. Further, through their research, conferences, and capacity-building programmes, both organizations are able to widely disseminate their ideas, thereby exerting considerable influence on how civil society, international organizations, the private sector, and even city governments view cities. This, in turn, has led to tangible outcomes such as public-private partnership initiatives, programmes, and policy mechanisms that create direct linkages between the local and international levels, channel resources to cities to enhance their position vis-à-vis national governments, and engage the city as a 'strategic partner' in global governance.

[140] AIMF, 'Qui sommes-nous?', online: www.aimf.asso.fr/default.asp?id=107 (accessed on 3 April 2016).
[141] AIMF, 'Liste des membres', online: www.aimf.asso.fr/default.asp?id=40 (accessed on 3 April 2016).
[142] AIMF, 'Programmes', online: www.aimf.asso.fr/default.asp?id=13 (accessed on 3 April 2016).

3.3.1 *Liveable, Competitive, Well Governed, and Bankable*

The World Bank, one of the most powerful multilateral development institutions in the world, has been involved in urban development since the 1970s.[143] With an annual loan budget of about US$15–20 million and a budget of US$25 million for research alone, the World Bank has the financial means to put its ideology into practice.[144] Furthermore, the Bank has throughout its history consistently managed to 'appropriate key aspects of the debates, inflect them to suit its own agendas, and endorse its positions such that they become … the official, conventional, or commonsense views such that everyone else follows suit until a new debate arises'.[145] Thus, the World Bank's representations on cities and urbanization have hegemonic influence in the international community; they create an ideological platform that promotes linkages between cities, globalization, and development, and, in some ways, support for the rise of cities in international affairs.

Since the 1970s, the World Bank's urban agenda has shifted its emphasis from project-based lending (to improve living conditions in slum settlements) to macroeconomic management in the 1980s and 1990s and eventually towards privatization. At the core of Ramsamy's remarkable study, 'The World Bank and Urban Development', lies the argument that the shifts in the Bank's urban policy were not purely the results of technocratic decision-making based on technical evaluations of projects, but rather responses to geopolitical and intellectual trends both within and outside the Bank.[146] Currently, in spite of references to civil society, sustainability, and 'pro poor policies', the Bank's urban policy retains the conservative orientation associated with neo-liberalism and market fundamentalism.[147]

The Bank's latest urban and local government strategy, published in 2009, and its earlier strategy, published in 2000, contain many neo-liberal policy recommendations. In its *Cities in Transition* strategy (2000), the bank set out its vision of sustainable cities. According to this vision, in order to be sustainable and functional, a city must be '*competitive, well governed and managed*, and financially sustainable, or *bankable*'.[148] The strategy defines 'bankability' as 'financial soundness in the

[143] Edward Ramsamy, *World Bank and Urban Development: From Projects to Policy* (Routledge 2006), pgs. 2–3.

[144] The World Bank is the largest sources of official finance for urban development; Christine Kessides, *Cities in Transition: World Bank Urban and Local Government Strategy* (Washington, DC: World Bank, 2000), pg. 58.

[145] Ramsamy, pg. 78. [146] Ramsamy, pg. 3.

[147] I prefer to avoid using the term 'Washington consensus' because its popular contemporary usage to connote a dogmatic commitment to the belief that the market offers the solution to every policy issue is a wide departure from its original meaning. For discussion, see John Williamson, 'What Should the World Bank Think about the Washington Consensus?' (2000) 15 *World Bank Research Observer* 251; John Williamson, *A Short History of the Washington Consensus* (Paper commissioned by Fundación CIDOB for conference 'From the Washington Consensus Towards a New Global Governance', Barcelona, September 24–25, 2004), online: https://piie.com/sites/default/files/publications/papers/williamson0904-2.pdf (accessed on 3 May 2016).

[148] Kessides, pg. 8.

treatment of revenue sources and expenditures – and, for some cities, a level of creditworthiness permitting access to the capital market'.[149] It goes on to state that '[f]or potentially creditworthy cities, the Bank's urban assistance should be geared to helping them access the capital market … Creative and flexible forms of Bank Group support will be especially important where central governments, often wisely, do not wish to continue providing sovereign guarantees to subnational governments after decentralization."[150]

Since 2000, the World Bank has been promoting a suite of measures to help local governments gain access to international financial markets.[151] The Bank launched a major programme, 'Capital Markets at the Sub-National Level', in 2000 to provide officials with technical assistance to improve their cities' capacity to access capital markets and held a major conference in New York City on sub-national government financing.[152] In 2013, the World Bank and the Public-Private Infrastructure Advisory Facility (PPIAF), a trust fund that is housed within the World Bank, launched the City Creditworthiness Academy to work with developing cities to improve their credit ratings and secure private-sector financing for infrastructure projects.[153] Interestingly, the City Creditworthiness Academy is part of the World Bank Group's Low Carbon, Livable Cities initiative, which focuses on improving cities' planning and financing capabilities so that they can better implement low-carbon development strategies. The concept of bankability has therefore been deployed to enhance cities' ability to mitigate and adapt to climate change as well.

On the one hand, it can be argued that these initiatives to improve cities' creditworthiness and access to capital markets empower cities and increase their independence from higher levels of government. From a macro perspective, these initiatives constitute processes that facilitate the disaggregation of the state and the rescaling of global politics, thereby creating space for cities to increase their involvement in international affairs. On the other hand, and from a less optimistic viewpoint, the World Bank's urban strategy will primarily affect the future of cities in developing countries, which are the primary recipients of World Bank technical and financial assistance. This raises the concern that the World Bank will destroy the diversity of cities in the quest to create social conditions that facilitate global capitalism.[154] In this line of thinking, as cities increasingly become 'private cities' –

[149] Ibid., pg. 11. [150] Ibid., pg. 12.

[151] For comprehensive discussion of the theory and practice of sub-national lending, see Mila Freire et al. (eds), *Subnational Capital Markets in Developing Countries: From Theory to Practice* (World Bank and Oxford University Press 2004).

[152] Details of the conference, *World Bank: Global Conference on Capital Markets Development at the Subnational Level* (New York City, 15–18 February 2000), can be found here: www.ce-review.org/00/1/wb1_factsheet.html (accessed on 1 May 2016).

[153] City Creditworthiness Initiative: A Partnership to Deliver Municipal Finance: online: www.worldbank.org/en/topic/urbandevelopment/brief/city-creditworthiness-initiative (accessed on 1 May 2016).

[154] For discussion, see Frug and Barron; Janne E. Nijman, '*The Future of the City and the International Law of the Future*' in Sam Muller et al. (eds), *The Law of the Future and the Future of Law* (Torkel Opsahl Academic EPublisher [Oslo, Norway] 2011), pgs. 217–218.

that is, '[cities] that envision city power principally as a mechanism for promoting private economic development'[155] – they create a network of sub-national authorities and spaces that primarily serve the needs of the transnational capitalist class at the risk of disenfranchising the less well-off sectors of urban society.[156]

The World Bank has also been a staunch advocate of decentralization, which has been promoted as an important aspect of the 'good governance' paradigm. In its current manifestation, 'good governance' has come to refer to the institutional conditions that will enable a well-functioning market for goods and services to emerge.[157] It is strongly associated with the neo-liberal commitment to free markets and privatization. As the concentration of power and resources in the hands of central governments is seen as a significant source of market distortion (for example, because of bureaucratic red tape and corruption), decentralization of power and authority is seen as an essential step towards achieving good governance.[158] At the same time, cities are perceived to be the level of government that is closest to the people.[159] The assumption is that decentralization will bring decision-making closer to local communities, create bureaucracies that are locally accountable and therefore more responsive to the needs of the people, and create effective ways to counter corruption.[160] This thinking on the part of international organizations happens to coincide nicely with the campaign by cities for more autonomy. Basing their arguments on the potential for cities to foster democracy and on the claim that city government is the level of government closest to the people and therefore most responsive to their needs, organizations like UCLG have been advocating for greater autonomy for cities as well as more decentralization.[161] The nexus of values and interests has led to cities and international organizations cooperating to promote decentralization and subsidiarity (a principle borrowed from EU law).[162] One of the consequences has been the embrace by international organizations of the city as a

155 Frug and Barron, pg. 4.

156 The transnational capitalist class is described as 'comprised of the owners of transnational capital, that is, the group that owns the leading worldwide means of production as embodied principally in the transnational corporations and private financial institutions'; B. S. Chimni, 'International Institutions Today: An Imperial Global State in the Making' (2004) 15 *European Journal of International Law* 1, pg. 4.

157 Porras, pg. 553. 158 Porras, pg. 554.

159 See, for example, Kessides, pg. 35; Muhammad Amjad Saqib, 'Introduction' in Syed Mubashir Ali and Muhammad Amjad Saqib (eds), *Devolution and Governance: Reforms in Pakistan* (Oxford University Press 2008), pg. 1.

160 Saqib, ibid. Also see UN Development Programme, *Decentralised Governance for Development: A Combined Practice Note on Decentralisation, Local Governance and Urban/Rural Development* (April 2004), online: www.undp.org/content/dam/aplaws/publication/en/publications/democratic-governance/dg-publications-for-website/decentralised-governance-for-development-a-combined-practice-note-on-decentralisation-local-governance-and-urban-rural-development/DLGUD_PN_English.pdf (accessed on 20 May 2016).

161 UCLG, 'About Us', online: www.uclg.org/en/organisation/about (accessed on 20 May 2016).

162 On the subsidiarity principle in EU law, see Antonio Estella, *The EU Principle of Subsidiarity and Its Critique* (Oxford University Press 2002).

'partner' and an alternative interlocutor to the state. This lays the foundation for cities to claim and assert political authority in the global order.

3.3.2 *Make Cities and Human Settlements Inclusive, Safe, Resilient, and Sustainable*

The UN system officially recognized the challenges of urbanization for the first time when the first Conference on Human Settlements was held in Vancouver in 1976.[163] This conference (Habitat I) resulted in the creation of the UN Commission on Human Settlements and the UN Centre for Human Settlements (which served as the secretariat of the Commission).[164] From 1978 to 1996, these two entities, 'with meager financial and political support', struggled to address the problems of rapid urbanization, especially in the developing world.[165] The second Conference on Human Settlements (Habitat II) took place in Istanbul in 1996, resulting in the Habitat Agenda, which contained over 100 commitments and 600 recommendations.[166] The concept of sustainable development is at the heart of the Habitat Agenda, which UN-Habitat was tasked to implement. Eventually, in 2002, the UN General Assembly adopted a resolution that transformed the Commission and its Centre for Human Settlements into what is now known as the UN Human Settlements Programme, or UN-Habitat. This resolution also recognized UN-Habitat as a subsidiary organ of the General Assembly[167] and the focal point for all human settlement matters within the UN system.[168] UN-Habitat's mandate is to promote socially and environmentally sustainable towns and cities.[169]

As a result of the Conference on Housing and Sustainable Urban Development (Habitat III) that took place in Quito, Ecuador, in October 2016, there has been a renewed global commitment to the concept of sustainable urbanization.[170] Led by UN-Habitat and driven by a range of partners including NGOs, private foundations, local authorities, and national governments, the World Urban Campaign was a global platform 'to promote dialogue, sharing, and learning about our urban future'

[163] UN Habitat, 'About Us'. [164] Ibid. [165] Ibid.

[166] United Nations Conference on Human Settlements (Habitat II), *The Habitat Agenda* (UN Doc A/Conf195/14, 7 August 1996).

[167] Para. A1, UN General Assembly, 56/206 *Strengthening the Mandate and Status of the Commission on Human Settlements and the Status, Role and Functions of the United Nations Centre for Human Settlements (Habitat)* (UN General Assembly 56th Session, Agenda Item 102, A/RES/56/206, 26 February 2002).

[168] Para B1, ibid.

[169] UN-Habitat's mandate is set out in various UN documents, including Resolution 56/206 and the Istanbul Declaration on Human Settlements (Habitat II) and the Habitat Agenda. For details, see UN-Habitat, 'Mandate and Role within the UN System', online: http://unhabitat.org/about-us/history-mandate-role-in-the-un-system/ (accessed on 1 May 2016).

[170] Habitat III, 'About Habitat III', online: www.habitat3.org/the-new-urban-agenda (accessed on 1 May 2016).

ahead of the Habitat III conference.[171] In its Manifesto for Cities, the Campaign called on the international community to recognize that the current models of urbanization are socially, environmentally, and economically unsustainable and that a new paradigm is needed to achieve a more sustainable future.[172] This message was reinforced in the World Cities Report 2016, the global flagship report on sustainable urban development launched by UN-Habitat in May 2016.[173]

Sustainable urbanization envisions 'equitable, resilient, livable, creative and productive cities' and incorporates measures for poverty reduction, environmental preservation, and good governance.[174] In addition, sustainable urbanization calls for the reduction of the ecological footprint of cities through 'integrated and holistic urban development policies, effective and participatory planning and management ... and the use of methodologies and tools to track urban sustainable development'.[175] It is believed that the promotion of the rule of law is essential to the success of sustainable urban development. In this regard, rule of law in cities – otherwise referred to as urban law in the policy literature – includes promoting accountability and transparency as well as participation by citizens in public decision-making processes.[176] This in turn will help secure the rights of city dwellers and the key principles of the New Urban Agenda.

The New Urban Agenda is the outcome document agreed upon at the Habitat III conference.[177] The implementation of the New Urban Agenda is envisioned to contribute to the 'implementation and localization' of the 2030 Agenda for Sustainable Development – known as the Sustainable Development Goals – and particularly Goal 11 (which is set out as the heading of this section).[178] The Agenda is

[171] UN-Habitat, World Urban Campaign, online: http://unhabitat.org/urban-initiatives/world-urban-campaign/ (accessed on 1 May 2016).

[172] *Manifesto for Cities: The Urban Future We Want*, pg. 5.

[173] UN-Habitat, 'UN-Habitat Launches the World Cities Report 2016, Urbanization and Development: Emerging Futures', online: http://unhabitat.org/un-habitat-launches-the-world-cities-report-2016/ (accessed on 1 May 2016).

[174] *Manifesto for Cities: The Urban Future We Want*, pg. 6. [175] Ibid.

[176] Urban law is defined as 'the collection of policies, laws, decisions and practices that govern the management and development of the urban environment'; UN-H1. Another useful definition of urban law is that it refers to a 'an expansive discipline that considers a range of traditional legal questions – local government authority, judicial review of regulatory processes and individual rights, among others – as they inform the life of cities'; Nestor M. Davidson, 'What Is Urban Law Today? An Introductory Essay in Honor of the Fortieth Anniversary of the Fordham Urban Law Journal' (2013) 40 *Fordham Urban Law Journal* 1579, pg. 1588. Davidson also notes that, in the United States, urban law has faded as a discipline in recent years because much of the subject matter that would be considered urban has shifted to other areas such as planning law, criminal justice and tax law.

[177] United Nations Conference on Housing and Sustainable Urban Development (Habitat III), *Draft Outcome Document of the United Nations Conference on Housing and Sustainable Urban Development (Habitat III)*, Doc. A/CONF.226/4*, 29 September 2016, online: https://habitat3.org/the-new-urban-agenda (accessed on 25 November 2016). Annexed to this document is the New Urban Agenda: Quito Declaration on Sustainable Cities and Human Settlements for All.

[178] Ibid., para. 9, Preamble of the Quito Declaration; UN, 'Transforming Our World: The 2030 Agenda for Sustainable Development', Sustainable Development Goals (SDGs), Goal 11, online: https://sustainabledevelopment.un.org/sdg11 (accessed on 1 May 2016).

underpinned by three principles. The first is to '[l]eave no one behind', which entails addressing social inequality, poverty and providing equitable access to physical and social infrastructure in the city.[179] The second principle is to achieve sustainable economic development by enhancing productivity, innovation, and competitiveness.[180] The third principle is to foster ecological and resilient human settlements by, amongst a number of things, protecting biodiversity and addressing climate change.[181] The agenda affirms the need for strengthened urban governance, 'with sound institutions and mechanisms to empower and include urban stakeholders'.[182] It also calls for the strengthening of municipal finance and local fiscal systems, and supports the creation of legal and regulatory frameworks to facilitate municipal borrowing from public and private sources.[183] Mechanisms to support expanded borrowing by city authorities include developing municipal debt markets and establishing regional, national, and sub-national development funds.[184]

On addressing climate change, the New Urban Agenda sets out some points that are worth noting. Paragraph 75 states a commitment 'to [encouraging] national, subnational and local governments ... to develop sustainable, renewable and affordable energy, energy-efficient buildings and construction modes; and to promoting energy conservation and efficiency, which are essential to enable the reduction of greenhouse gas and black carbon emissions'.[185] Paragraph 79 sets out an explicit commitment in support of the Paris Agreement. It states that signatories to the Quito Declaration commit themselves 'to promoting international, national, subnational and local climate action, including climate change adaptation and mitigation, and to supporting the efforts of cities and human settlements ... to be important implementers. We further commit ourselves to supporting building resilience and reducing emissions of greenhouse gases ... Such measures should be consistent with the goals of the Paris Agreement adopted under the United Nations Framework Convention on Climate Change, including holding the increase in the global average temperature to well below 2°C above pre-industrial levels and pursuing efforts to limit the temperature increase to 1.5°C above pre-industrial levels.'[186]

The New Urban Agenda, like the earlier Habitat Agenda, will provide guidance to states, city governments, civil society, private foundations, and international organizations in their thinking about cities, urbanization, and sustainable development. There are a number of observations to be made about the UN-Habitat and the New Urban Agenda in relation to the emergence of cities in international affairs. The Habitat III process has increased the visibility and prominence of cities not just as sites of urbanization processes but also as strategic partners in the quest to achieve the global development agenda. Since the late 2000s, international organizations,

[179] Para. 14(a), Quito Declaration. [180] Para. 14(b), Quito Declaration.
[181] Para. 14(c), Quito Declaration. [182] Para. 15(c)(ii), Quito Declaration.
[183] Para. 139, Quito Declaration. [184] Ibid. [185] Para. 75, Quito Declaration.
[186] Para. 79, Quito Declaration.

private foundations, and other actors have created multiple platforms and programmes to work with city governments on a range of issues.[187] In March 2004, the UN Educational, Scientific and Cultural Organization (UNESCO) launched an initiative known as the International Coalition of Cities against Racism.[188] This initiative established a network of cities interested in sharing information and experience in order to improve their anti-discrimination policies.

The description of the International Coalition of Cities against Racism on UNESCO's website is noteworthy because of what it reveals about the official thinking on the governance potential of cities. The website states as follows:

> The international conventions, recommendations or declarations elaborated at the upstream level need to be ratified and implemented by the States. At the same time, it is extremely important to involve actors on the ground including the targets of discriminations, to make sure that those instruments are applied to respond to concrete problems. *UNESCO chose cities as the privileged space to link upstream and downstream actions.* The role of city authorities as policy-makers at the local level, is considered here as the key to create dynamic synergies.[189]

By bypassing national governments, UNESCO is seeking more effective and responsive domestic implementation of international law. Once the New Urban Agenda is underway, it can be argued that initiatives similar to the International Coalition of Cities against Racism will expand in number and scope, creating yet more space and opportunities for cities to assert their role in global development and international affairs.

As mentioned earlier, the New Urban Agenda affirms a commitment to decentralization, subsidiarity, and local self-governance. The Agenda also calls for cities to work with the private sector and improve their fiscal systems. By reaffirming commitment to these ideas, the New Urban Agenda throws normative weight behind the policy discourse about harnessing the benefits and minimizing the ills of urbanization and, by extension, the indispensable role that cities will play in the global quest for sustainable urban development. In other words, UN-Habitat's ideology on sustainable urbanization very closely resembles the World Bank's urban strategy.

[187] While there has been a surge of interest on the part of UN agencies, the EU, and the World Bank in cities since the 2000s, it ought to be clarified that there are partnerships between international organizations and cities that predate this recent surge of interest. For example, the World Health Organization's (WHO) European Healthy Cities network has been in operation for over twenty-five years. In its first phase of implementation (1987–1992), the network was designed to serve as a social laboratory for testing health initiatives at the local level and providing feedback to WHO and states; Michele Acuto, Mika Morissette, and Agis Tsouros, 'City Diplomacy: Towards More Strategic Networking? Learning with WHO Healthy Cities' (2016) Global Policy DOI: 10.1111/1758-5899.12382, pg. 3.

[188] UNESCO, International Coalition of Cities against Racism, online: www.unesco.org/new/en/social-and-human-sciences/themes/fight-against-discrimination/coalition-of-cities/ (accessed on 1 May 2016).

[189] Ibid.

Although UN-Habitat tries to pay more attention to poverty alleviation and securing human rights such as the right to clean water and the right to shelter, there is striking consensus between the New Urban Agenda and the Bank's strategy.

3.4 CONCLUSION

This chapter concludes the first part of this book, which aims to provide a firm foundation for exploring the role of cities in governing climate change. This chapter drew from a wide variety of empirical data and secondary literature to make the claim that cities are on the rise in international affairs. As globalization and urbanization continue apace, cities have sought to play a role in global governance processes. Often, cities are motivated to do so because they are at the frontline of global governance challenges such as terrorism and pandemics. In some cases, cities seek to engage in diplomacy to facilitate trade and cultural exchange and to make pronouncements on issues such as LGBT rights.

As for the urban policies of international organizations, we saw in this chapter the pivotal role that the World Bank plays in shaping the development of cities and the policy discourse about the role of cities in the quest for sustainable development. Through its technical assistance, lending policies, and research, the World Bank is able to exercise significant influence on the international community's thinking about cities, globalization, and development. The Bank advocates decentralization and the principle of subsidiarity as part of its 'good governance' paradigm for development; cities are seen as the level of government that is closest to the people and therefore most suited for facilitating democratic participation in grassroots decision-making processes and developing mechanisms for accountability. In contrast, central governments are seen as distant bureaucratic machineries that are detached from the lives of the people and therefore unable to respond to their needs and aspirations. Consequently, the World Bank and other international organizations are increasingly bypassing the state and working directly with cities to achieve global governance objectives such as climate mitigation and countering discrimination. It is in this context that cities have the space in which they can engage directly in governance partnerships involving civil society, multinational corporations, and international organizations. It is also in this context that cities are proactively taking measures to respond to climate change and forming networks to share best practices. Through these networks, cities are also developing norms, practices, and voluntary standards that transcend national boundaries, to steer municipal governments and other actors towards climate change mitigation and low-carbon development in the cities of today and the future.

4

City Action on Climate Change

4.1 INTRODUCTION

The discussion in Chapter 3 showed how global cities are beginning to exert their presence in international affairs through acts of inter-city diplomacy and implementation of international law on their own accord. Global cities are also beginning to emerge as governance actors, and it is in the area of climate change that cities have been particularly active. To date, over 2,000 cities have specific plans to scale up their efforts at climate adaptation as well as strategies for GHG emission reductions.[1] These cities do not act alone. They work in partnership with venture capitalists, research laboratories, and universities, as well as with other cities.[2] Urban partnerships are often both local and transnational, and blur the private-public divide by involving private actors and public institutions alike.

This chapter showcases what five global cities – London, Mexico City, New York City, Rotterdam, and Seoul – are doing to address climate change. These global cities have consciously styled themselves as 'climate frontrunners'. Through their proactive climate change policies and strategies, they have gained widespread recognition from policymakers, urban planners, the media, international organizations, and researchers. The aim of this chapter is not to provide detailed case studies or a comprehensive account of each city's mitigation and adaptation efforts. It also does not purport to analyze how and why various cities differ in their laws and

[1] United Nations Climate Summit 2014, 'Mayors at UN Climate Summit Announce Pledges towards Major Carbon Cuts in Cities', online: www.un.org/climatechange/summit/2014/09/mayors-un-climate-summit-announce-pledges-towards-major-carbon-cuts-cities/ (accessed on 15 November 2016).

[2] For example, Clean Tech Delta is a Dutch initiative that brings together the city governments of Rotterdam and Delft, the business sector, research institutions, and laboratories to support the development of clean technologies in the Delft–Rotterdam–Drechtsteden region. Clean Tech Delta describes itself as a 'triple helix organization', referring to the concept of the triad of university–industry–government in the knowledge society; Interview No. 8. For a concise overview of the triple-helix concept, see Stanford University, 'The Triple Helix Concept', online: http://triplehelix.stanford.edu/3helix_concept (accessed on 15 June 2016).

policies.[3] Instead, this chapter aims to provide an overview that will give readers a sense of what global cities are doing 'on the ground' within their territories to govern climate change.[4]

An appreciation of the localized practices and practical aspects of global city action on climate change is important in at least two respects. One of the key ideas this book seeks to advance is that the practices, policies, and strategies taken at the local level bear wider significance and contribute towards transnational climate change governance when global cities seek to scale up their actions and pursue cooperation through cross-border networks. In this chapter, one observes the intertwined connections between urban institutions and global organizations, between local practices and transnational norms. In some cases, there are causal connections between practices at the local and global levels, which brings to mind Koh's concept of the transnational legal process whereby norms and practices are 'uploaded' and 'downloaded' from the international to lower levels of governance and vice versa. Further, it can be observed that a number of commonalities unite these climate frontrunner global cities. The features these cities share include visionary leadership by a mayor who is able to secure 'buy-in' from the city's administrative agencies, the private sector, civil society, and other major stakeholders; public support for strong climate action and, more broadly, creating a more sustainable and liveable home; and active participation in global networks such as C40, which is the focus of Chapter 5.[5] This helps put into perspective the role global cities can play in the transnational climate change regime complex. While many cities, large and small, may seek to contribute to the global effort to mitigate climate change, only a select few global cities with the money, administrative resources, political will, and cosmopolitan vision of the common good will be active participants in transnational governance and rule-making processes.

3 Why a city does or does not implement an initiative or a policy is often dictated by local politics and other highly localized factors. As one of the city government officials I interviewed puts it, 'Cities are political entities at the end of the day. They have to respond to local political demands and local electoral cycles. Municipal governments cannot implement policies if they do not have the support of the key stakeholders'; Interview No. 1. Interview No. 9 gave the example of the recent New York City Council's approval to levy a five-cent (US currency) levy on plastic bags. That New York City has been relatively late in introducing a tax to reduce the use of plastic bags 'is not for the lack of trying. We have faced a lot of political push-back'. The *Wall Street Journal* also reported that the deliberations on this five-cent fee levy in the New York City Council was unusually heated, and the '28–20 vote to approve the bill came only after a fierce debate centering on lofty themes of regressive taxation, income inequality and environmental policy'; Mara Gay, 'New York City Council Approves 5-Cent Fee on Plastic Bags' *Wall Street Journal* (6 May 2016).

4 Thus, a conscious choice has been made to highlight each city's most notable climate strategies, policies, and regulations rather than seek consistency in the type of information provided for each global city featured in this chapter.

5 For discussion on the motivational effects of city participation in networks, see Taedong Lee and Chris Koski, 'Mitigating Global Warming in Global Cities: Comparing Participation and Climate Change Policies of C40 Cities' (2014) 16(5) *Journal of Comparative Policy Analysis: Research and Practice* 475, pg. 490.

4.2 LONDON

With a population of 8.4 million in mid-2013, London is the most populous city in Europe, drawing vast numbers of people from the rest of the UK and the world.[6] London's economy, heavily dominated by the financial sector, contributes 20 per cent to the country's national output (measured by gross value added, or GVA).[7] London is also widely considered to be one of the world's great cities: it is 'the city of Empire, the most multicultural city in the world, a centre of financial globalization'.[8] At the same time, London has a sizable carbon footprint: the city emits as much GHGs as Greece or Portugal,[9] which brings home the point that effective climate policies to reduce the GHG emissions of a city can have as much impact as addressing those of a country.

Seizing upon its resources and the desire to play a leadership role in addressing climate change, London has set the high water mark in urban efforts to mitigate and adapt to climate change. Bulkeley and Schroeder argue that London's climate policy is more advanced and well developed vis-à-vis other global cities because of the following factors:

> The drivers and motivations ... are necessarily multiple and complex, but include the commitment of critical individuals, the courage of conviction born in part from interim policy success, a positive climate of public opinion, a lack of overt opposition from key interest groups and the emergence of new market opportunities in the carbon economy.[10]

Those 'critical individuals' include the former mayor, Ken Livingstone, and his deputy mayor, Nicky Gavron (2000 to 2008).[11] They made addressing the causes of climate change one of the main priorities of their mayoralty and set ambitious targets

[6] Office for National Statistics (United Kingdom), Population Estimates for the UK (2013), online: www.ons.gov.uk/ons/rel/pop-estimate/population-estimates-for-uk–england-and-wales–scotland-and-northern-ireland/2013/sty-population-estimates.html (accessed on 1 October 2016); European Commission, Eurostat, 'Statistics on European Cities', online: http://epp.eurostat.ec.europa.eu/statistics_explained/index.php/Statistics_on_European_cities (accessed on 3 October 2016).

[7] Office for National Statistics (United Kingdom), 'London's Economy Has Outperformed Other Regions since 2007', online: www.ons.gov.uk/ons/rel/regional-trends/regional-economic-indicators/march-2013/sum-london.html (accessed on 3 October 2016); for discussion on whether having a global financial centre like London boosts or harms the UK economy, see Howard Davies, 'Does London's Financial Centre Boost or Harm the UK Economy?', *Guardian* (25 February 2014), online: www.theguardian.com/business/economics-blog/2014/feb/25/london-financial-centre-boost-or-harm-uk-economy (accessed on 3 October 2016).

[8] Jennifer Robinson, 'Making London, through Other Cities' in Sarah Bell and J. Paskins (eds), *Imagining the Future City: London 2062* (London: Ubiquity Press 2013), pg. 24.

[9] London Climate Change Agency, *Moving London towards a Sustainable Low-Carbon City: An Implementation Strategy* (2007), pg. 1.

[10] H. Bulkeley and H. Schroeder, *Governing Climate Change Post-2012: The Role of Global Cities – London* (Tyndall Centre for Climate Change Research Working Paper 123, 2008), pg. 8.

[11] Ibid., pg. 10; Heleen Lydeke, P. Mees, and Peter P. J. Driessen, 'Adaptation to Climate Change in Urban Areas: Climate-Greening London, Rotterdam, and Toronto' (2011) 2 *Climate Law* 251, pg. 271. It should be noted that London has a two-tier government structure: the Greater London Authority is

and policies for both mitigation and adaptation.[12] The C40 network is also the brainchild of former mayor Livingstone. Since its inception, the network has had multiplier effects across the globe and has also brought London significant economic benefits and extended its 'soft power'. Although C40 will be the subject of detailed discussion in Chapter 5, the salient point for present purposes is to note the pivotal role that London, particularly its former mayor, played in establishing C40.

The following section will first discuss London's mitigation efforts led by the strategic vision of Mayor Livingstone, which continues to exercise considerable influence today. London's mitigation efforts focus on retrofitting the city's existing buildings to be more energy efficient, promoting renewable energy and, in the long term, transiting to an economy powered by hydrogen as a low-carbon energy source. It then examines London's adaptation initiatives, which are largely led by the boroughs because of their jurisdictional control over spatial planning.

4.2.1 London's Mitigation Policies and Programmes

During his first term, then-mayor Livingstone developed an Energy Strategy for London.[13] Climate change was at the heart of this strategy, which committed London to reducing carbon dioxide emissions by 20 per cent below 1990 levels by 2010 as a first step in a reduction of 60 per cent by 2050.[14] This strategy focused on promoting the use of on-site renewable energy generation (for example, by the use of solar panels) and combined heat and power (CHP). To promote renewable energy or 'decentralized energy generation', Livingstone used his powers in the planning system – which included approving large-scale developments and devising the London Plan, the plan that sets the overarching framework for spatial development across the London boroughs – to ensure that new developments included decentralized energy generation.[15] The Mayor's Office and the London Development Agency also publicly backed high-profile projects that demonstrate the technical feasibility of renewable energy and CHP. Livingstone's office also developed multistakeholder partnerships focused on adaptation (the London Climate Change Partnership) and research and development for new hydrogen technologies (the

the regional body that consists of the mayor and an assembly that provides oversight of the mayor's work, and there are thirty-three boroughs or local authorities.

[12] Ibid.; also see John Vidal, 'Ken Livingstone, the Mayor of London, Is on a Mission to Tackle Climate Change' *Guardian* (1 November 2006), online: www.theguardian.com/environment/2006/nov/01/travelsenvironmentalimpact.localgovernment (accessed on 3 October 2016).

[13] Greater London Authority, *Green Light to Clean Power: The Mayor's Energy Strategy* (Greater London Authority 2004).

[14] Ibid.

[15] The Greater London Authority Act bestowed the mayor and the London Assembly with a range of new powers that expanded the mayor's ability to direct local planning authorities. For discussion, see Christopher Stanwell, 'Devil Lies in the Detail as Mayor Increases Powers', Planning Resource, online: www.planningresource.co.uk/article/767412/legal-report (accessed on 6 October 2016).

London Hydrogen Partnership, established in 2002).[16] Schroeder and Bulkeley describe the period from 2000 to 2004 as 'one of experimentation with the formal powers of the mayor in relation to energy and transport policy, and the emergence of a partnership approach to climate governance in London'.[17]

The period of 2004 to 2008 was marked by even more concerted efforts to galvanize climate change action in London. The momentum continued apace with the establishment of the London Climate Change Agency to deliver Livingstone's Energy Strategy by implementing projects in the sectors that impact climate change, especially in the energy, transport, waste, and water sectors.[18] The 2007 London Climate Change Action Plan set out in greater detail how technical and regulatory barriers to promoting renewable energy, energy efficiency in commercial and residential buildings, and development of a hydrogen energy infrastructure would be overcome. The 2007 London Climate Change Action Plan also established a more ambitious policy goal: 'to stabilize carbon dioxide emissions in 2025 at 60% below 1990 levels, with steady progress towards this over the next twenty years'.[19] However, it was recognized in the Action Plan that the 'difficult truth is that in preparing this action plan we have been unable to present any realistic scenario in which we can achieve the 2025 target set out above, without major national regulatory and policy change'.[20]

To ensure the continuity of London's climate-focused development agenda, the Greater London Authority Act 2007 imposed a new statutory duty on the mayor of London to contribute towards adaptation and mitigation.[21] In fulfilment of this duty, the mayor is required to produce statutory mitigation and adaptation strategies for London.[22] Thus, while some of the above-mentioned drivers or motivations may alter because of changes in leadership or economic conditions, the legal requirement to address climate change constitutes a firm substantive basis upon which future climate policy for London can be built. Following the removal of Livingstone and Gavron from office after the May 2008 elections and the appointment of a mayor from the Conservative Party, Boris Johnson, it was unclear whether the new mayor would place the same degree of emphasis on climate change and whether London will continue to be a global leader on climate change. Developments from 2008 to 2014 indicated that Mayor Johnson did not detract from the commitments made in the London Climate Change Action Plan. The policies and strategies envisioned and pursued by the Livingstone mayoralty were still implemented, even though Mayor Johnson was not personally committed to addressing climate change

[16] London Climate Change Partnership website: http://climatelondon.org.uk/lccp/; London Hydrogen Partnership website: www.hydrogenlondon.org (accessed on 6 December 2016).

[17] H. Schroeder and H. Bulkeley, 'Global Cities and the Governance of Climate Change: What Is the Role of Law in Cities ?' 36 *Fordham Urban Law Journal* 313, pg. 335.

[18] London Climate Change Agency, pg. 3.

[19] Greater London Authority, *Action Today to Protect Tomorrow: The Mayor's Climate Change Action Plan* (2007), pg. 19.

[20] Ibid. [21] Section 42, Greater London Authority Act 2007. [22] Sections 43 and 44, ibid.

the way his predecessor was and there was a tendency by Johnson's office to emphasize the 'business case' for addressing climate change.[23]

In May 2016, Sadiq Khan won the election contest and replaced Boris Johnson as mayor of London.[24] Within the first few months of Khan's mayoralty, it was already evident that climate change would become a more central concern than it had been for the previous mayor. Keen to work with other global cities to address climate change and for London to revive its role as a climate frontrunner amongst European cities, Sadiq Khan increased contact with C40 and put himself up as a candidate for the role of vice chair on the C40 Steering Committee.[25] On 25 July 2016, C40 announced that Sadiq Khan had been elected to the vice chair position and will represent European cities alongside the mayor of Copenhagen.[26] In addition, although Sadiq Khan's office had yet to publish its climate change strategy for the city at the time of writing, the mayor has already made it clear that he wants to establish London as a 'low-carbon beacon' and will commit London to become a zero-carbon city by 2050.[27] Finally, it is expected that Khan will support more integrated approaches such as enhancing the synergies between climate adaptation, expanding green spaces, and reducing air pollution.[28]

4.2.2 *Adaptation: Implementation at the Local Authority Level*

In pursuance of its obligations under the Greater London Authority Act 2007, the Greater London Authority issued the Draft Climate Change Adaptation Strategy for London (Adaptation Strategy) in 2008 (followed by an updated version in 2010).

[23] See Greater London Authority, *The Mayor's Climate Change Mitigation and Energy Annual Report (with updated carbon dioxide emissions for 2011 and updated activity for 2011–2013)* (Greater London Authority 2014). For evidence of personal commitment to addressing climate change, see Boris Johnson, 'It's Snowing, and It Really Feels Like the Start of a Mini Ice Age', *Telegraph* (20 January 2013), online: www.telegraph.co.uk/comment/columnists/borisjohnson/9814618/Its-snowing-and-it-really-feels-like-the-start-of-a-mini-ice-age.html (accessed on 8 October 2014). For emphasis on the 'business case' of tackling climate change, see the second objective of Mayor Johnson's climate change strategy: 'The global market for low carbon goods and services is around £3 trillion. In addition to this, it is estimated that if global CO_2 emission targets are met, this market could increase by at least £368 billion per year through to 2030. London is well placed to capitalize on this economic opportunity. The Mayor is using London's inherent strengths and his climate change mitigation programmes to build on this'; Greater London Authority, 'Delivering London's Energy Future: The Mayor's Climate Change Mitigation and Energy Strategy ', Executive Summary, pg. viii.

[24] Robert Booth, 'Labour's Sadiq Khan Elected Mayor of London', *Guardian* (7 May 2016), online: www.theguardian.com/politics/2016/may/07/sadiq-khan-elected-mayor-of-london-labour (accessed on 10 December 2016).

[25] The Steering Committee is the governing body that provides strategic direction to the C40 network.

[26] C40, 'Press Release: London Mayor Sadiq Khan Elected C40 Vice Chair', online: www.c40.org/press_releases/press-release-london-mayor-sadiq-khan-elected-c40-vice-chair (accessed on 1 November 2016).

[27] Sadiq Khan for London, online: www.sadiq.london/a_greener_cleaner_london (accessed on 1 November 2016).

[28] Interview No. 12.

Mees and Driessen note that '[a]lthough flood risk is a major driver of adaptation, heat stress has become more important after the heat waves of 2003 and 2006'.[29] This reality is also reflected in the Adaptation Strategy, which highlights flooding, drought, and overheating as the three main climate change impacts that London has to manage.

A significant prong of London's Adaptation Strategy consists of an Urban Greening Programme to reduce the impacts of and prevent further intensification of London's urban heat island effect. The Urban Greening Programmes consists of a green roof policy, a tree-planting programme, and establishment of a green grid across the Greater London area.[30] In the Adaptation Strategy, Mayor Johnson committed to working with partners 'to increase green cover in central London by five per cent by 2030 and a further five per cent by 2050, to manage temperatures in the hottest part of London ... increase tree cover across London by five per cent (from 20 to 25 per cent) by 2025 ... [and] enable the delivery of 100,000 m2 of new green roofs by 2012 (from 2008/09 baseline)'.[31] Here, partners include the boroughs that play a critical role in developing and implementing locality-specific adaptation policies and initiatives under the broad framework established at the mayoral level.

At the local authority level, some boroughs such as Barking and Dagenham have issued a Planning Advice Note on Green Roofs.[32] The borough of Barking and Dagenham has also issued a Planning Advice Note on Sustainable Design and Construction which 'strongly [encourages developers] to consider energy efficient building design, community heating and CHP, as well as renewable energy generation from the outset of their plans'.[33] Further, the Planning Advice Note states as follows: 'Developers in Barking and Dagenham are expected to achieve a further 20 per cent reduction in carbon emissions beyond Building Regulations 2006 in new developments. Ten per cent of this reduction has to be achieved through on-site

29 Mees and Driessen, pg. 261.

30 See London Climate Change Partnership, *Adapting to Climate Change: Lessons for London* (Greater London Authority 2006), for detailed consideration of case studies that informed London's adaptation strategy – for example, Linz's and Basel's established green roof policies; see City of London Corporation, *Green Roof Case Studies* (City of London Corporation 2011), online: www.cityoflondon.gov.uk/services/environment-and-planning/planning/heritage-and-design/Documents/Green-roof-case-studies-28Nov11.pdf (accessed on 7 October 2016) for successful examples of London's fledgling green roofs initiative.

31 Greater London Authority, *The Draft Climate Change Adaptation Strategy for London* (Greater London Authority 2010), pg. 12, online: www.london.gov.uk/sites/default/files/Draft_Climate_Change_Adaptation_Strategy.pdf (accessed on 10 October 2016).

32 Planning Advice Notes provide developers and planners guidance on how a borough seeks to implement its Local Development Framework. While they are not binding and do not have to be adhered to in applications for planning permission, the borough will take into account whether these notes have been followed when it decides on a planning application; London Borough of Barking and Dagenham, 'Planning Advice Notes', online: www.lbbd.gov.uk/Environment/PlanningPolicy/Pages/Planningadvicenotes.aspx (accessed on 10 October 2016).

33 London Borough of Barking and Dagenham, Sustainable Design and Construction (Planning Advice Note 5), pg. 24, online: www.lbbd.gov.uk/Environment/PlanningPolicy/Documents/PAN5SustainableDesign.pdf (accessed on 10 October 2016).

generation of renewable energy. The remainder 10 per cent can be achieved through increased energy efficiency, CHP or through further generation of renewable energy.'[34] Thus, at the local authority level, there is discretion for boroughs to be ambitious in their climate change agenda and go beyond statutory requirements.

London is vulnerable to a number of flood risks: from the North Sea (tidal flooding), the Thames river and its tributaries (fluvial flooding), and from heavy rainfall (surface water flooding).[35] The Draft Adaptation Strategy assessed that '[c]urrently, there is a low likelihood of tidal flooding, a medium probability of river flooding and high probability of surface water flooding'.[36] However, the risks will increase as sea levels rise, tidal surges increase in height, and the amount and intensity of winter rainfall increases. Currently, flood risk in London is managed primarily by a system of flood defences (walls, gates, and the Thames Barrier) and drainage networks. However, spatial planning is also a critical adaptation tool, which involves avoiding the creation of flood-vulnerable land uses in high-risk areas and using the planning process to reduce flood risks. As the London boroughs are the primary decision-makers in the planning process, they have prepared a range of spatial planning policies targeted at adaptation to flooding. The borough of Sutton, for example, has adopted a holistic 'green-blue infrastructure' approach to manage flood risks to and from new developments and to promote sustainable urban drainage systems to manage surface water flooding.[37] This firm focus on creating and maintaining green and blue spaces was the result of Sutton's participation in the Green and Blue Space Adaptation for Urban Areas and Eco Towns (GRaBS) project, a network of fourteen partners drawn from eight EU member states that aims to share experience and best practices in integrating adaptation into regional planning and development.[38]

4.3 MEXICO CITY

With nearly 20 million inhabitants, 3.5 trillion vehicles, and 35,000 industrial facilities, Mexico City's per capita carbon emissions is estimated at 3.6 tonnes, which is low compared to those of wealthier cities like Los Angeles, which has per capita emissions of 15.6 tonnes.[39] Mexico City's carbon emissions represent

[34] Ibid., pg. 27.

[35] Greater London Authority, *The Draft Climate Change Adaptation Strategy for London*, pg. 7.

[36] Ibid.

[37] See Patrick Whitter, 'London Borough of Sutton: Adaptation to Flooding via Local Planning Policies' in Aleksandra Kazmierczak and Jeremy Carter (eds), *Adaptation to Climate Change Using Green and Blue Infrastructure: A Database of Case Studies* [database prepared for the Interreg IVC Green and blue space adaptation for urban areas and eco towns (GRaBS) project 2010], online: www.grabs-eu.org /membersArea/files/Database_Final_no_hyperlinks.pdf (accessed on 8 October 2016).

[38] The GRaBS project is financed by the European Union's Regional Development Fund. More information is available online: www.grabs-eu.org (accessed on 8 October 2016).

[39] Patricia Romero Lankao, 'How Do Local Governments in Mexico City Manage Global Warming?' (2007) 12 *Local Environment* 519, pg. 520.

4.18 per cent of the total national emissions – a relatively small percentage because of the absence of the petrochemical industry in the city (one of the main sources of GHGs in Mexico) and increasing deindustrialization of the city.[40] However, Mexico City, like other urban centres in Latin America and the Caribbean (LAC) is a growing source of GHG emissions because of urban sprawl and population growth.[41] Mexico City also suffers from some of the worst air pollution problems in the world.[42] In 1992, the United Nations reported that Mexico City was the most polluted city on the planet.[43]

Studies of the impacts of climate change across Mexico indicate that Mexico City is likely to experience the greatest effects because of its existing environmental problems (induced by rapid urbanization) and its susceptibility to climatic events.[44] Historically, Mexico City has experienced catastrophic floods and severe droughts. Projected changes in temperature and precipitation may increase the frequency and severity of floods and droughts, increasing the city's susceptibility to rain and wastewater flooding and landslides as well as dramatically reducing residents' access to clean drinking water. Furthermore, large-scale rapid urban expansion, coupled with changes in atmospheric emissions, has resulted in temperature rises, high levels of air pollution, and loss of ecosystem services for Mexico City. Since the early 1990s, the city's average temperature has risen by almost four degrees Celsius.[45] The city also faces significant challenges meeting the water, sanitation, and housing demands of its population, of which some 14 per cent live in slum settlements in flood-prone areas.[46] Despite its inland location, Mexico City is prone to flooding because it has no natural drainage outlet and the city is located on the marshy bed of what was once

[40] Fabiola S. Sosa-Rodríguez, 'From Federal to City Mitigation and Adaptation: Climate Change Policy in Mexico City' (2014) 19 *Mitigation and Adaptation Strategies for Global Change* 969, pg. 989. As people move away from the core areas of the city to suburban areas, they have to travel longer distances for work and recreational purposes. Private automobile ownership has been steadily increasing because of the poor public transportation system in Mexico City. In LAC, emissions from transportation are the fastest-growing source of carbon. Mexico City is no exception.

[41] Jorgelina Hardoy and Patricia Romero Lankao, 'Latin American Cities and Climate Change: Challenges and Options to Mitigation and Adaptation Responses' (2011) 3 *Current Opinion in Environmental Sustainability* 158.

[42] Galen McKinley et al., *The Local Benefits of Global Air Pollution Control in Mexico City: Final Report of the Second Phase of the Integrated Environmental Strategies Program in Mexico* (Instituto Nacional de Ecología, México, Instituto Nacional de Salud Pública, México, 2003), pg. 6.

[43] World Health Organization and United Nations Environment Programme, *Urban Air Pollution in Megacities of the World* (Blackwell, Oxford 1992), pg. 39.

[44] Patricia Romero Lankao et al., 'Institutional Capacity for Climate Change Responses: An Examination of Construction and Pathways in Mexico City and Santiago' (2013) 31 *Environment and Planning C: Government and Policy* 785, pg. 789.

[45] Sosa-Rodríguez, pg. 972.

[46] Romero Lankao et al., pg. 790 (Table 2). There is also wide income disparity between the rich and the poor in Mexico City, which has a GINI coefficient of 0.56.

a series of lakes in a valley.[47] The poor and marginalized are the most vulnerable to the impacts of climate change.

These challenges have rendered it imperative for Mexico City to increase its climate resilience. In recent years, Mexico City has emerged as a 'climate frontrunner' amongst global cities in developing countries.[48] For many developing country cities which generally emit less GHGs compared to wealthy cities in developed countries, '[c]urbing carbon emissions may ... not be the current "local environmental priority". Rather, the local priority may be vulnerability and adaptation to the impacts of climate change.'[49] This has not been the case for Mexico City. Unusually for an emerging economy, domestic climate policy discussions in Mexico and its capital, Mexico City, have not focused solely on impacts and adaptation but have also included mitigation responsibilities.[50] This is all the more remarkable because Mexico City, like many major cities in rapidly emerging economies, faces tremendous challenges such as high levels of migration from rural areas into the city as it contemplates a transition towards a more climate-friendly and sustainable future.[51]

The discussion that follows will first set out Mexico City's climate policies as well as explain the main drivers behind the city's ambitions to address climate change. These factors include (1) the coincidence of a presidency (for the country) and a mayoralty (for the city) being held by individuals committed to addressing environmental and climate change issues; (2) Mexico City's involvement in transnational networks, particularly C40; and (3) the availability of external funding from international organizations like the World Bank and the Kyoto Protocol's Clean Development Mechanism (CDM). The section concludes with a brief discussion of the Programa para Mejorar la Calidad del Aire en el Valle de México (ProAire), which has led to impressive reductions in conventional air pollution and carbon emissions in Mexico City.

47 Priscilla Connolly, 'The Case of Mexico City, Mexico' (case study report prepared for Understanding Slums: Case Studies for the Global Urban Report 2003), online: www.ucl.ac.uk/dpu-projects/Global_Report/home.htm (accessed on 6 November 2016).

48 The term 'developing country' is used to refer to low- and middle-income economies. Mexico is classified as an upper-middle-income economy [gross national income (GNI) of US$4,126 to US$12,745] because its GNI per capita stands at US$9,940 (2003 data); World Bank, online: http://data.worldbank.org/country/mexico (accessed on 6 November 2016).

49 Romero Lankao, pg. 520.

50 Simone Pulver, 'A Climate Leader? The Politics and Practice of Climate Governance in Mexico' in David Held, Charles Roger, and Eva- Maria Nag (eds), *Climate Governance in the Developing World* (Wiley 2013), pg. 175. Mexico is the only developing country to have enshrined long-term GHG emissions reduction targets in national legislation and, in fact, is only the second country in the world to do so (the first being the United Kingdom); Richard Black, 'Inside Mexico's Climate Revolution' *BBC News* (20 April 2012), online: www.bbc.com/news/science-environment-17777327 (accessed on 4 November 2016).

51 Interview No. 10.

4.3.1 *Strong Mayoral Commitment to Tackling Climate Change*

From 2000 to 2005, Andrés Manuel López Obrador served as the mayor of Mexico City. Described as a visionary, Mayor Obrador 'did not need convincing that climate change was a serious problem and one that Mexico City needed to address'.[52] Under Mayor Obrador's leadership, Mexico City produced its first Local Climate Action Strategy in 2004, establishing guidelines for local governmental agencies, the private sector, and civil society to promote mitigation and adaptation.[53] The mitigation measures focused on improving Mexico City's public transportation system, improving energy efficiency, and increasing green acreage in the city. To build the city's climate resilience, the strategy identified that the immediate priority was to increase public education of risks such as heat stroke and enhancing early warning systems.

Although Mayor Obrador was responsible for placing climate change on the city's agenda and putting the institutional framework in place, it was during Marcelo Ebrard's mayoralty (2006–2012) that Mexico City displayed a high level of commitment and invested significant resources to developing and implementing an ambitious range of policies and programmes to curb the city's GHG emissions and increase its climate resilience. Ebrard was committed to improving the quality of life for his city's 20 million inhabitants, and that included reducing drug-related violence, improving access to education, and improving the city's environment.[54] A crucial component of Mayor Ebrard's 'green city' strategy was to realize the co-benefits of reducing Mexico City's severe air pollution and realizing GHG emissions reductions at the same time.[55] As transportation and fossil fuel–based energy production are key sources of carbon emissions and conventional air pollutants (such as nitrogen oxide and sulphur dioxide), a climate mitigation strategy will usually produce the co-benefit of improved air quality.[56]

There are two key factors that motivated Mayor Marcelo Ebrard to undertake ambitious climate action. The first was Mexico City's participation in transnational climate networks, particularly C40 and the World Mayors Council on Climate Change.[57] C40 first emerged in 2005 and rapidly built a reputation for being an elite group of leading global cities that are strongly committed to tackling climate change.[58] Mexico City was keen to be part of C40 and to prove its credentials as a progressive global city that is pursuing ambitious climate action. C40's

52 Ibid.

53 Secretaría del Medio Ambiente del Gobierno del Distrito Federal (Ministry of Environment of the Federal District), *Estrategia Local de Acción Climática de la Ciudad de México*, 2004.

54 Joel Jaeger, 'Reflecting on Marcelo Ebrard's Tenure as the Mayor of Mexico City' Council on Hemispheric Affairs (Mexico, 20 September 2012) front page, online: www.coha.org/reflecting-on-marcelo-ebrards-tenure-as-the-mayor-of-mexico-city/ (accessed on 4 November 2016).

55 Interview No. 10. On the co-benefit approach, see Netherlands Environmental Assessment Agency (PBL), *Co-Benefits of Climate Policy* (PBL Report No. 500116005, 2009), online: www.unep.org/transport/gfei/autotool/understanding_the_problem/Netherlands%2520Environment%2520Agency.pdf (accessed on 4 November 2016).

56 Ibid. 57 Interview No. 10. 58 See detailed discussion of C40 in Chapter 5.

membership criteria are widely known to be demanding, and the quest to meet those membership requirements motivated Mayor Ebrard's administration to pursue an ambitious climate change agenda. Once Mexico City officially became a C40 member, it can be said that frequent interactions with other cities and participation in the network provided a healthy dose of peer pressure that added impetus to Mayor Ebrard's climate change agenda.[59]

As previously mentioned, Mayor Ebrard was also active in another transnational network known as the World Mayors Council on Climate Change.[60] The mayor of Kyoto, Yorikane Masumoto, initiated the founding of the World Mayors Council on Climate Change soon after the Kyoto Protocol entered into force in February 2005.[61] This network of mayors seeks to represent and advocate for cities on matters pertaining to global sustainability.[62] Mayor Ebrard was not only an active participant; in 2009, he was elected to be the chairman of the council, and Mexico City hosted the 2010 World Mayors Summit on Climate Change.[63] This was a proud moment for Mexico City and its mayor. The summit resulted in the Global Cities Covenant on Climate, otherwise known as the Mexico City Pact.[64] More than 250 cities in fifty-seven countries have signed the pledge to reduce GHG emissions, undertake adaptation measures, and furthermore to record 'their climate actions … and provide regular information and data so that [their] efforts can be measured, reported and verified'.[65]

The second factor is the coincidence in timing between Marcelo Ebrard's mayoralty and Felipe Calderón's presidency of the country. When Felipe Calderón assumed office in 2006, it marked a turning point in the national discourse on climate change. President Calderón made climate change a focus of his presidency. His administration produced a National Strategy for Climate Change outlining the various mitigation and adaptation options that would be taken to achieve climate resilience as well as the long-term goal of a 50 per cent reduction in GHGs by 2050.[66] This coincidence helped create synergistic energy between Mexico City's climate governance ambitions and the national agenda.[67] For example, the national government was supportive of Mexico City's ambitious climate change programmes

[59] Ibid.
[60] World Mayors Council on Climate Change, online: www.worldmayorscouncil.org (accessed on 8 November 2016).
[61] Ibid. [62] Ibid.
[63] Local Government Climate Roadmap, 'World Mayors Summit on Climate – Mayors Push for Hope after Copenhagen', online: www.iclei.org/climate-roadmap/advocacy/global-lg-events/2010-world-mayors-summit-on-climate-mexico-city.html (accessed on 8 November 2016).
[64] Global Cities Covenant on Climate 'Mexico City Pact', online: www.mexicocitypact.org/downloads/texto-original/Global%2520Cities%2520Covenant%2520on%2520Climate%2520OFICIAL.pdf (accessed on 8 November 2016).
[65] Ibid. The carbon*n* Climate Registry is a public online database that has been established to support cities in their voluntary adherence to Article 4; online: http://carbonn.org/ (accessed on 8 November 2016).
[66] Pulver, pg. 184. [67] Interview No. 10.

because it was open to the idea of the capital city serving as a social laboratory to test policies and initiatives that can be scaled up to the national level. During the 2006–2012 period, this coincidence in domestic politics meant that Mexico City was well placed to take ambitious local climate change action as well as participate actively in transnational climate change governance through C40 and the World Mayors Council on Climate Change.

In 2008, Mexico City was the first municipality in LAC to implement a Local Climate Action Program (2008–2012). In this programme, the city set itself two objectives: first, to reduce carbon dioxide equivalent (CO_2e) emissions by 7 million tonnes during 2008 to 2012, and secondly, to initiate an integrated and fully functional climate adaptation programme by 2012.[68] This programme consisted of twenty-six mitigation strategies, which included programmes to replace the city's streetlamps with energy-efficient light-emitting diode (LED) lamps; increasing the use of solar energy in hospitals and government buildings; building the Ecobus system to transport 150,000 passengers daily, as well as nine Metrobus (bus rapid transit) systems and Line 12 of the Metro (subway) to transport 437,000 passengers daily; and introducing more sustainable waste management policies. The capture and use of biogas at the Bordo Poniente Stage 4 landfill represented nearly 90 per cent of the emission reductions in the waste sector, while the construction of Line 12 of the city's subway system and an obligatory school transportation programme (whereby educational institutions require their students to take public transportation to school) are estimated to be responsible for half the emission reductions in the transport sector.

The twelve adaptation strategies included improving watershed management; building flood protection infrastructure in urban ravines; instituting soil remediation programmes to improve the city's natural rainwater absorption capacity; and providing social assistance to vulnerable social groups, such as by distribution of free food and establishing shelters for the homeless during heatwaves.[69] Long-term adaptation measures are mainly geared towards developing adaptive capacity in rural areas that face severe marginalization by promoting organic farming, development, and natural resource conservation.[70] It should be noted that the Local Climate Action Program (2008–2012) integrates strategies from the city's General Program of Development (2006–2012) and the Green Plan (2006–2012).[71] These strategies may not have been developed with climate change in mind, but they assist in reducing the city's GHG emissions and increasing its adaptive capacity. In 2010, Mexico City passed the Climate Change Mitigation and Adaptation Law which authorizes the city's government to take certain regulatory measures for mitigation and adaptation, authorizes the city's government to regulate actions for addressing

[68] Ministry of the Environment and Federal Government of the District of Mexico City, *Mexico City Climate Action Program 2008–2012 (Summary)* (2008), pg. 11, online: www.planningclimatechange.org/joomla/o_upload/PDF_unico.pdf (accessed on 6 November 2016). The information in the rest of this paragraph is taken from this document, unless otherwise indicated.

[69] Ibid., pg. 18. [70] Ibid. [71] Ibid., pg. 4.

climate change, and establishes the Climate Change Environmental Fund to provide funding for initiatives such as the establishment and management of GHG emission inventories and the creation of a carbon emissions trading system.[72]

In 2012, Mayor Marcelo Ebrard announced that not only had Mexico City met its goal of reducing CO_2e emissions by 7 million tonnes during 2008 to 2012, but that it had surpassed this goal by 10 per cent, with a 7.7 million tonne reduction of CO_2e emissions over the four-year period.[73] Mayor Ebrard also announced that Mexico City had met all the adaptation objectives laid out in the Local Climate Action Program.[74] However, academic commentators have been less positive in their assessment of Mexico City's achievements. Sosa-Rodríguez, for example, points out that there is limited inter-institutional coordination and collaboration because climate mitigation and adaptation is viewed as the responsibility of the Ministry of the Environment.[75] Further, some national adaptation strategies such as implementing drought alert systems have yet to be introduced in Mexico City. Sosa-Rodríguez explains that the '[i]dentified obstacles to successful [mitigation and adaptation] in the city include a lack of understanding of the strategies' objectives, process and outcomes by governmental agencies and inhabitants, . . . a lack of participation and public awareness about climate change'.[76] These problems have resulted in poor coordination and collaboration among these participants to address climate change impacts. It should also be noted that after the current mayor of Mexico City, Miguel Mancera, assumed office in 2012, he significantly restructured the city's administration, and this contributed to confusion and lack of coordination amongst various agencies that should be working together to implement the city's climate change strategy.[77] Furthermore, problems of corruption have stood in the way of developing the city's mitigation and adaptation capacity.[78] These institutional and societal barriers to action tend to be more common in cities located in the developing world than those in developed countries and reflect broader socioeconomic and governance challenges. Thus, unlike cities such as Rotterdam, London, and New York which enjoy more stable legal and political institutions to support climate action, a city like Mexico City has emerged as a climate frontrunner in the face of far greater social and political resistance.

4.3.2 *The ProAire Programmes*

Air pollution has been a critical issue in Mexico City, and the authorities continue to face an uphill battle in improving the city's air quality.[79] Since the 1990s, a series of

[72] Romero Lankao et al., pg. 792.
[73] C40 News Team, 'Mexico City Meets, Exceeds Climate Action Program Goals' *National Geographic* (13 September 2012), Voices, online: http://voices.nationalgeographic.com/2012/09/13/mexico-city-meets-exceeds-climate-action-program-goals/ (accessed on 4 November 2016).
[74] Ibid. [75] Sosa-Rodríguez, pg. 985. [76] Ibid. [77] Interview No. 10.
[78] Sosa-Rodríguez, pg. 986.
[79] David Agren, 'Mexico City Chokes on Its Congestion Problem' *Guardian* (6 July 2016).

comprehensive air quality improvement programmes have improved air quality and in recent years have also served as a platform for reducing GHG emissions. The first plan, Programa Integral para el Control de la Contaminación Atmosférica (PICCA) was initiated in 1990 and met with considerable success in introducing two-way catalytic converters, phasing out the use of leaded gasoline, and establishing vehicle emissions standards.[80] The second programme, ProAire II (1995–2000) introduced restrictions on the aromatic content of fuels and reduction of sulphur content in industrial fuel.[81] Notwithstanding significant improvements in ambient air quality, levels of conventional air pollutants still far exceeded World Health Organisation (WHO) guidelines, compelling the city government to extend the ProAire program.[82] While air quality remained the key focus of the ProAire III (2002–2010) programme, there was greater recognition of the linkages between air pollution, urbanization, transportation, and climate change.[83]

The ProAire III programme includes eighty-nine control measures, which range from closing the city's most polluting factories to banning cars one day per week in the city's metropolitan area. Its Metrobus (bus rapid transit) system, launched in 2005, is the longest such system in Latin America. It is estimated that, by introducing cleaner, more efficient buses, and convincing many commuters to leave their cars at home, Metrobus has reduced carbon dioxide emissions from Mexico City traffic by an estimated 80,000 tonnes a year.[84] The new fleet of buses, operating on clean-burning ultra low-sulphur diesel fuel, makes more than 450,000 trips per day.[85] In 2009, the Mexico City Metrobus System was awarded the Roy Family Award for Environmental Partnerships by the John F. Kennedy School of Government at Harvard University. The award recognizes outstanding public-private partnership projects that enhance environmental quality through the use of novel and creative approaches.[86] The city's Ecobici bike-sharing programme is also the largest in the LAC and has been replicated in other cities in the region. The ProAire IV programme (2011–2020) contains measures across eight strategy areas including energy consumption, greening of the municipal transport fleets, education, green roofing and reforestation, capacity building, and scientific research.[87] In 2013, the long-term approach and success of the ProAire

[80] McKinley et al., pg. 6. [81] Ibid. [82] Ibid. [83] Romero Lankao, pg. 525.

[84] Belfer Center for Science and International Affairs, *Harvard Kennedy School's Belfer Center Announces 2009 Roy Family Award for Environmental Partnership*, online: http://belfercenter.ksg.harvard.edu/publication/19541/harvard_kennedy_schools_belfer_center_announces_2009_roy_family_award_for_environmental_partnership.html (accessed on 8 November 2016).

[85] Ibid.

[86] Ibid. The Metrobus system is a result of a partnership launched by the World Resources Institute Center for Sustainable Transport, together with CEIBA (a Mexican NGO) and the Mexico City government. The project received funding from Shell Foundation, Caterpillar Foundation, Hewlett Foundation, and the World Bank.

[87] City Climate Leadership Awards, '2013 – The Winners', online: http://cityclimateleadershipawards.com/2013-city-climate-leadership-awards-winners/ (accessed on 8 November 2016).

programmes garnered Mexico City the City Climate Leadership Award in the air quality category.[88]

4.4 NEW YORK CITY

New York City (NYC) is the largest city in the United States and a global financial centre.[89] The city has a population of about 8.4 million (as of July 2013) that is estimated to reach 9 million in 2040.[90] NYC's high population density of over 10,000 people per square kilometre, extensive public transit system, and dominance of the financial sector in its local economy shape its GHG emission patterns. In general, NYC's total emissions are high, but its per capita emissions are relatively low compared to other urban areas in the United States.[91] NYC's total GHG emissions were estimated to be 61.5 million metric tonnes of CO_2e in 2007.[92] The city's per capita emissions were estimated to be 7.1 metric tonnes of CO_2e – higher than London's estimated 5.9 metric tonnes, but lower than estimates for San Diego (11.1) and San Francisco (11.2).[93] A key reason for NYC's relatively low emissions compared to other American cities is the prevalent use of public transport amongst NYC's residents. Across the United States, transportation is the second-largest source of GHG emissions, accounting for 28 per cent of total emissions.[94] NYC is the only US city where more than 50 per cent of the population does not drive to work, and it has the highest rate across the country of commuting by public transit.[95] NYC's GHG emissions profile is comprised largely of energy-related CO_2 emissions since there is little agricultural or forested land within the city and 75 per cent of the methane produced at the city's landfills and wastewater treatment plants is captured.[96] NYC's first GHG emissions inventory, completed in 2007, showed that more than two-thirds

[88] Ibid.

[89] NYC usually takes second place (after London) by a narrow margin in the Global Financial Centres Index; see Long Finance's 'Financial Centres Futures' Program, 'The Global Financial Centres Index 13', March 2013, online: www.geneve-finance.ch/sites/default/files/pdf/2013_gfci_25march.pdf (accessed on 9 November 2016).

[90] NYC Department of City Planning, 'Current Estimates of New York City's Population for July 2013', online: www.nyc.gov/html/dcp/html/census/popcur.shtml (accessed on 11 November 2016); NYC Department of City Planning, 'New York City Population Projections by Age/Sex & Borough, 2010–2040' online: www.nyc.gov/html/dcp/pdf/census/projections_report_2010_2040.pdf, pg. 2 (accessed on 11 November 2016).

[91] Lily Parshall et al., *The Contribution of Urban Areas to Climate Change: New York City Case Study* (case study prepared for Cities and Climate Change: Global Report on Human Settlements 2011), pg. 6.

[92] City of New York, 'Inventory of New York City Greenhouse Gas Emissions (2007)', online: www.nyc.gov/html/planyc2030/downloads/pdf/emissions_inventory.pdf (accessed on 11 November 2016).

[93] Parshall et al., pg. 7.

[94] United States Environmental Protection Agency (US EPA), 'National Greenhouse Gas Emissions Data', online: www.epa.gov/climatechange/ghgemissions/usinventoryreport.html (accessed on 11 November 2016).

[95] Parshall et al., pg. 7.

[96] Ibid.

of the city's emissions are due to electricity consumption in residential, commercial, and institutional buildings.

Hurricane Sandy, the most destructive hurricane of the 2012 Atlantic hurricane season, caused severe flooding, power cuts, and forty-four deaths in NYC. While it took 'an improbable set of factors coming together to give rise to the catastrophic effects of the storm',[97] Hurricane Sandy represents a fraction of the climate risks that NYC faces. The NYC Panel on Climate Change has projected that, by 2050, sea levels could rise up to thirty inches.[98] This will pose a significant threat to NYC's many low-lying neighbourhoods during storms and tidal flooding. The panel also predicts that, by 2050, there will be an increase in the most intense hurricanes occurring in the North Atlantic basin and heatwaves could triple in frequency because of the city's densely built environment and the urban heat island effect that causes temperatures in NYC to be up to seven degrees (Fahrenheit) higher than in surrounding areas. According to the NYC Panel on Climate Change, there is a 90 per cent probability that NYC will experience more frequent heavy downpours. The discussion that follows sets out the policies and measures that have been implemented since NYC's then-mayor Michael Bloomberg decided to take concerted action on climate change in his second term (2005–2013). PLANYC is a comprehensive plan outlining the mayor's vision for a more sustainable city and, for the first time, has set a GHG emissions reduction goal for NYC. Since the launch of PLANYC, the city has passed more than a hundred laws and regulations to address climate change.[99] As has been observed about the other cities discussed in this chapter, a crucial element behind a city's ability to undertake ambitious climate action is high-level political leadership. NYC is no exception, with Mayor Bloomberg acting as 'the champion of the climate change issue for the city, guiding the overall process with great foresight and courage'.[100]

Succeeding Michael Bloomberg in 2014, the current mayor of NYC is Bill de Blasio. Mayor de Blasio campaigned on a platform focused on tackling NYC's growing income and social inequality.[101] Therefore, his agenda so far, while

97 PLANYC, 'The Risks We Face: Sandy and Its Impacts', online: www.nyc.gov/html/planyc/html/resiliency/sandy-impacts.shtml (accessed on 11 November 2016).

98 The information included in this sentence and the rest of this paragraph is drawn from PLANYC, 'The Risks We Face: Climate Change', online: www.nyc.gov/html/planyc/html/resiliency/climate-change.shtml (accessed on 11 November 2016). The New York City Panel on Climate Change produces updated, peer-reviewed local projections for NYC climate risks. The mayor's offices in charge of sustainability and resilience rely on the climate science provided by the New York City Panel on Climate Change, and there is frequent interaction between the mayor's offices and the panel; Interview No. 9.

99 New York City Council, *Comprehensive Platform to Combat Climate Change* (2014), online: http://council.nyc.gov/html/pr/climateagenda.pdf (accessed on 11 November 2016).

100 Katherine Bagley and Maria Gallucci, 'How Mayor Michael Bloomberg Thought Big on Climate' *Scientific American* (20 December 2013), online: www.scientificamerican.com/article/bloomberg-climate-plan-genesis-excerpt/ (accessed on 11 November 2016).

101 For discussion, see, for example, George Packer, 'Bill de Blasio's Vision' *The New Yorker* (12 August 2013).

demonstrating a very high level of commitment to mitigation and building the city's climate resilience, is built on a broad understanding of social, environmental, and economic sustainability; tackling climate change has become one of four core issues instead of the main focus of his mayoralty as it was for Michael Bloomberg.[102]

4.4.1 *PLANYC: A Sustainability and Climate Change Blueprint for the 'City That Never Sleeps'*

As early as the mid-1990s, policy experts and scientists were producing studies that warned of the climate vulnerabilities that NYC faced and how climate change would drive up risks such as heat-stress mortality and mosquito-borne diseases.[103] However, these warnings were ignored by the Giuliani mayoral administration (1994–2001), which took the position that 'that if action was indeed necessary, it could be delayed, because climate change was a long-term problem'.[104]

When the Bloomberg mayoral administration (2002–2013) took office, its first term was consumed with the aftermath of the terrorist attacks on 11 September 2001.[105] However, according to Bagley and Gallucci, Mayor Bloomberg' s interest in climate change started to increase in the mid-2000s; he was also concerned about how NYC was going to cope with 9 million residents in the near future, given that the densely populated city's transportation, housing, and public spaces were already under significant strain.[106] Bloomberg's administration, alongside over twenty-five city agencies, began the process of developing a strategy for how NYC was to prepare for its growing population, become more resilient, and reduce its carbon footprint. This process culminated in the publication of PLANYC.

Released in April 2007, PLANYC is a comprehensive programme of action comprising 127 initiatives in the key areas of land, water, transportation, energy, air quality, and climate change.[107] The key objectives of PLANYC are, inter alia, 'to create homes for almost a million more New Yorkers while making housing and neighbourhoods more affordable and sustainable, ensure all New Yorkers live within a ten minute walk of a park, and reduce energy consumption and make our energy systems cleaner and more reliable'.[108] In relation to climate change, the goal is a 30 per cent reduction of GHG emissions by 2030 from 2005 levels and to '[i]ncrease the resiliency of our communities, natural systems, and infrastructure to climate risks'.[109] A central objective of PLANYC was the establishment of an interagency

[102] See following discussion about Mayor de Blasio's climate agenda.

[103] Cynthia Rosenzweig and William Solecki, 'Chapter 1: New York City Adaptation in Context' (2010) 1196 *Annals of the New York Academy of Sciences* 19, pg. 20.

[104] Bagley and Gallucci, ibid. [105] Ibid. [106] Ibid.

[107] PLANYC, 'About PLANYC', online: www.nyc.gov/html/planyc/html/about/about.shtml (accessed on 11 November 2016).

[108] Ibid.

[109] Ibid. On 7 July 2008, Mayor Bloomberg announced a long-term plan to reduce energy consumption from municipal buildings and operations and a goal of reducing GHG emissions by 30 per cent by

taskforce that would identify climate risks and implement adaptation strategies across agencies.[110] Other objectives include updating the city's Federal Emergency Management Administration (FEMA) 100-year floodplain maps, documenting the city's floodplain management strategies to secure discounted flood insurance for New Yorkers, and amending the building code to address the impacts of climate change.

The MillionTreesNYC campaign is one of the PLANYC initiatives: it aims to plant a million trees in NYC by 2017. The campaign is a public-private partnership between the NYC Department of Parks & Recreation (NYC Parks) and a non-profit group, the New York Restoration Project.[111] Since its launch, public, private, and non-profit organizations have organized nearly 4,000 citizen volunteers to plant trees across the city.[112] One aspect of the MillionTreesNYC campaign directs the planting of nearly 400,000 trees to establish 2,000 acres of new forest on NYC parkland and other public open spaces. The aim is to create multilevel, ecologically functioning forests that will provide the city with numerous benefits such as filtering pollution from the local atmosphere, storing and sequestering carbon dioxide, and trapping rainwater during heavy storms. According to the PLANYC 2014 progress report, the MillionTreesNYC campaign is 27 per cent ahead of schedule, having planted over 830,000 trees (of over 120 species) and held numerous events to teach New Yorkers how to care for the trees.

PLANYC stands out for three reasons. First, it is a comprehensive strategy that integrates climate change, environmental protection and remediation, population planning, transportation management, and many other aspects that go towards making a city a truly sustainable one that is also enjoyable and exciting to live in. Many cities and states produce climate action plans, which as the name suggests focus on climate change and then seek to implement the initiatives by 'mainstreaming' them across various policy areas such as housing and transportation, usually with a degree of difficulty because the agencies responsible for these policy areas view climate change as the sole responsibility of the environmental agency.[113]

2017. It was estimated that this would reduce NYC's annual output of GHGs by nearly 1.7 million metric tonnes; Rosenzweig and Solecki, pg. 21.

110 Ibid.

111 Lisa Foderaro, 'As City Plants Trees, Some Say a Million Are Too Many' *New York Times* (18 October 2011), online: www.nytimes.com/2011/10/19/nyregion/new-york-planting-a-million-treestoo-many-some-say.html?pagewanted=all&_r=0 (accessed on 11 November 2014).

112 P. Timon McPhearson et al., 'Assessing the Effects of the Urban Forest Restoration Effort of MillionTreesNYC on the Structure and Functioning of New York City Ecosystems' (2010) 3 *Cities and the Environment* 1, pg. 2.

113 On the same point, see discussion in Edoardo Croci, Sabrina Melandri, and Tania Molteni, *A Comparative Analysis of Global City Policies in Climate Change Mitigation: London, New York, Milan, Mexico City and Bangkok* (Working Paper No. 32, Center for Research on Energy and Environmental Economics and Policy at Bocconi University, 2010), pg. 29. For an example of a city facing such difficulties, see discussion on Mexico City in this chapter. For discussion of the silo effect hindering climate action at the state or provincial level, see, for example, Jolene Lin, 'Climate

To a large extent, this 'silo effect' was avoided by getting all city agencies involved in the development of PLANYC – a process that would have been an important learning experience for the city servants who would bear the bulk of the responsibility for the strategy's eventual implementation. Further, then-mayor Bloomberg saw the importance of establishing a new office dedicated to overseeing the development of PLANYC. Established by Local Law 17 of 2008, the Mayor's Office of Long-Term Planning and Sustainability (OLTPS) works with all other city agencies to 'develop and coordinate the implementation of policies, programs and actions to meet the long-term needs of the city, with respect to its infrastructure, environment and overall sustainability citywide, including but not limited to the categories of housing, open space, brownfields, transportation, water quality and infrastructure, air quality, energy, and climate change' and track the progress of PLANYC.[114]

Secondly, NYC depended heavily on legislative action to implement PLANYC. More than a hundred new laws have been introduced and many more amended to update NYC's historic zoning regulations and address relatively novel issues such as electric car charging stations.[115] In relation to adaptation, the land use planning process and regulations in the Zoning Resolution (which is the city's zoning policy document) have been recognized as powerful tools to implement adaptation measures. For example, two amendments to the Zoning Resolution impose minimum requirements for landscaping and planting in yards and increase requirements for planting street trees.[116] The amendments aim to increase the city's vegetated and pervious surfaces to assist in stormwater management. On the mitigation front, an example would be the Greener, Greater Buildings Plan (GGBP) legislative package enacted in December 2009 to reduce energy consumption and increase energy efficiency.[117] The GGBP package consists of Local Law 85 (NYC Energy Conservation Code), Local Law 84 (Energy and Water Benchmarking), Local Law 87 (Energy Audits and Retro-commissioning) and Local Law 88 (Lighting Upgrades and Sub-metering).[118]

Governance in China: Using the "Iron Hand"' in Benjamin J. Richardson (ed), *Local Climate Change Law: Environmental Regulation in Cities and other Localities* (Edward Elgar 2012).

[114] New York City Council, Local Law 17 of 2008: Creation and Implementation of a Comprehensive Environmental Sustainability Action Plan for NYC, section 2.

[115] NYC enacted the first comprehensive zoning resolution in the country in 1916. It is therefore regarded as a pioneer in this field of zoning policy, which 'determines the types of uses permitted in different districts and the relationships among those districts'; Edna Sussman et al., 'Climate Change Adaptation: Fostering Progress through Law and Regulation' (2010) 18 *New York University Environmental Law Journal* 55, pg. 64.

[116] Ibid., pg. 66.

[117] PLANYC, 'Greener, Greater Buildings Plan', online: www.nyc.gov/html/gbee/html/plan/plan.shtml (accessed on 11 November 2016).

[118] 'Local laws that conflict with state statutes are expressly not authorized under the home rule powers of local government. Further, local laws that are authorized under the home rule powers may nevertheless be preempted if the state legislature chooses to occupy that particular field of regulation'; Sussman et al., pg. 132. Thus, the extent to which NYC can use local laws to implement PLANYC is subject to limitations imposed by federal and state law.

Thirdly, it is clear that vast amount of resources are being devoted to the implementation of PLANYC, a luxury that many other cities simply cannot afford. In April each year, a PLANYC progress report is published and made available to the public. Each progress report typically runs into more than a hundred pages and includes a highly detailed checklist and status update on the various initiatives. The 2014 PLANYC progress report, for example, provides an update on more than 400 specific milestones that include launching a consumer education campaign on flood insurance, working with pipeline operators to expand the city's natural gas supply, supporting the Health and Hospital's Corporation effort to protect public hospital emergency departments from flooding, fortifying all marinas and piers, and implementing economic revitalization programmes for areas devastated by Hurricane Sandy.[119] In addition, Local Law 17 of 2008 requires the OLTPS to issue an update to PLANYC every four years. The plan was updated in 2011 to include five more initiatives.[120] After Hurricane Sandy, more federal funding was made available to NYC to repair and restore the affected areas and, in the process, implement climate resilience measures and improve the energy efficiency of new buildings.[121]

4.4.2 *One New York: The Plan for a Strong and Just City*

Since Mayor Bill de Blasio took office, the OLTPS now comprises the Mayor's Office of Recovery and Resiliency (responsible for climate resilience) and the Office of Sustainability (responsible for the city's mitigation efforts). Following Hurricane Sandy, the Bloomberg administration had drawn up a detailed recovery plan known as *A Stronger, More Resilient New York*. This plan outlined a US$3.7 billion initial phase of thirty-seven coastal protection initiatives designed to protect vulnerable neighbourhoods and infrastructure from storm surge and sea level rise by increasing coastal edge elevations and improving coastal management.[122] Under Mayor de Blasio's leadership, PLANYC and *A Stronger, More Resilient New York* have been incorporated into the flagship strategy of his new administration, *One New York: The Plan for a Strong and Just City*.

As mentioned earlier, Mayor de Blasio's mayoral campaign focused on tackling inequality in NYC. In *One New York*, Mayor de Blasio states that this blueprint for the city's development will 'embrace equity as central to that work'.[123] *One*

[119] City of New York, *Progress Report 2014: A Greener, Greater New York/A Stronger, More Resilient New York* (2014), online: www.nyc.gov/html/planyc2030/downloads/pdf/140422_PlaNYCP-Report_FINAL_Web.pdf (accessed on 13 November 2016).

[120] Ibid., pg. 5.

[121] Interview No. 9; for information about the federal funds NYC received to repair and restore areas affected by Hurricane Sandy, see 'NYC Recovery: Community Development Block Grant Disaster Recovery', online: www.nyc.gov/html/cdbg/html/home/home.shtml (accessed on 13 June 2016).

[122] City of New York, pg. 58.

[123] *One New York: The Plan for a Strong and Just City*, online: www.nyc.gov/html/onenyc/downloads/pdf/publications/OneNYC.pdf, pg. 3 (accessed on 13 June 2016).

New York is built on four pillars: job growth/economic development, reducing social inequality, reducing the city's environmental and carbon footprint, and improving the city's resilience to the impacts of climate change.[124] NYC has set itself the goal of reducing its GHG emissions by 80 per cent (from a 2005 baseline) by 2050.[125] The *OneNYC 2016 Progress Report* shows that the city has been making good strides towards achieving the targets and implementing the numerous initiatives outlined in *One New York*.[126] There are concerns that the current mayoralty is trying to do too much and therefore losing the discipline to act concertedly on climate change.[127] However, at the same time, it is recognized that NYC is markedly ahead of many other cities as a climate frontrunner because of the early and ambitious start that the Bloomberg administration gave the city.[128] Furthermore, the Mayor's Office of Recovery and Resiliency and the Office of Sustainability presently constitute a full-time staff of around fifty people.[129] This represents a significant amount of human resources devoted to addressing climate change, which puts NYC in the enviable position of being able to continue developing and implementing ambitious and innovative urban climate governance solutions. Finally, during de Blasio's mayoralty, NYC continues to actively participate in C40 and 100 Resilient Cities (a network pioneered and financially supported by the Rockefeller Foundation).[130] Employees of the Mayor's Office regularly participate in C40 workshops to share NYC's experiences and best practices.[131] NYC's interactions with other global cities on issues of climate change take place primarily through these two networks, and such engagement is viewed as an important component of NYC's contribution towards global climate governance efforts.[132]

4.5 ROTTERDAM

For a city with a population size of about 620,000, Rotterdam has high emissions.[133] In 2014, the city emitted 30,414 kilotonnes of carbon dioxide,[134] making it one of the

[124] Ibid., pgs. 5–7.

[125] On 19 September 2014, just days ahead of the UN Climate Summit held in NYC on 23 September 2014, the NYC Council announced a comprehensive package of legislation and policies which, inter alia, set a new, more ambitious target of reducing the city's GHG emissions by 80 per cent (from a 2005 baseline) by 2050 and commits to passing new legislation to require the city to build 'zero carbon' buildings; New York City Council, *Comprehensive Platform to Combat Climate Change*. The GHG emissions reduction goal has been incorporated into *One New York: The Plan for a Strong and Just City*.

[126] OneNYC 2016 Progress Report, online: www1.nyc.gov/html/onenyc/downloads/pdf/publications/OneNYC-2016-Progress-Report.pdf (accessed on 13 June 2016).

[127] Interview No. 9. [128] Ibid. [129] Ibid. [130] Ibid. [131] Ibid. [132] Ibid.

[133] City of Rotterdam, Facts and Figures 2013, online: www.rotterdam.nl/Clusters/Stadsontwikkeling/Document%25202014/Informatiepunt%2520Arbeidsmarkt/ZigZag2013-Engels-DEF.pdf (accessed on 13 June 2016).

[134] Rotterdam Climate Initiative, CO2 Monitor 2014, version 1.0 (10 June 2015), pg. 1 (on file with author).

highest carbon dioxide emitting cities in Europe.[135] Within the national context, the Rijnmond economy (Rijnmond is the conurbation surrounding Rotterdam) is responsible for 8.5 per cent of the Netherland's gross domestic product (GDP), while also generating about 18 per cent of the country's total carbon dioxide emissions.[136] A significant portion of Rotterdam's emissions are port-related – which is not surprising, given that Rotterdam's port is the largest in Europe, with a total cargo throughput of 430 million tonnes in 2010.[137] Using vessel movements to estimate shipping-related emissions in ports, it has been shown that shipping-related emissions in Rotterdam represent 10 per cent of the shipping emissions in all European ports.[138] Efforts to manage Rotterdam's port-related GHG emissions can therefore have a significant global impact, a realization not lost on Rotterdam's city government which spearheaded the World Ports Climate Initiative.

The following section will first describe the city of Rotterdam's climate mitigation initiatives; this is followed by an account of the port of Rotterdam's climate mitigation efforts. With regard to port-related efforts to address climate change, particular emphasis will be placed on Rotterdam's founding role in the World Ports Climate Initiative, a voluntary effort undertaken by fifty-five of the world's largest ports to individually and jointly work together to reduce the climate impacts of port operations and the global shipping industry. The focus then shifts to Rotterdam's innovative and ambitious adaptation efforts that have earned it the reputation as a leading, if not *the* leading, 'climate-proof' delta city.

A common thread that runs through all of Rotterdam's climate initiatives is the significant role that C40 participation played in motivating Rotterdam to take ambitious climate change action and, furthermore, to take a leadership role on the global stage in the areas of port-related mitigation and building resilience in delta cities. It should also be noted that in 2013, Rotterdam was selected to participate in the 100 Resilient Cities programme, pioneered by the Rockefeller Foundation to empower cities to develop resilience.[139] Resilience is understood as '[t]he capacity of individuals, communities, institutions, businesses, and systems within a city to survive, adapt, and grow no matter what kinds of chronic stresses and acute shocks they experience'.[140] Becoming a member of this programme marked a turning point

135 Daniel Hoornweg, Lorraine Sugar, and Claudia Lorena Trejos Gomez, 'Cities and Greenhouse Gas Emissions: Moving Forward' (2011) 20 *Environment and Urbanization* 1, pg. 6.

136 City of Rotterdam, *Investing in Sustainable Growth: Rotterdam Programme on Sustainability and Climate Change 2010–2014* (City of Rotterdam), pg. 25.

137 Olaf Merk and Theo Notteboom, *The Competitiveness of Global Port-Cities: The Case of Rotterdam/Amsterdam – the Netherlands* (OECD Regional Development Working Papers 2013/06, 2013), pg. 21. Rotterdam is also the largest port for Switzerland, the second largest in Austria, and is an important port for central European countries such as the Slovak Republic, Hungary, and the Czech Republic; ibid., pg. 31.

138 Ibid. pg. 61.

139 Gemeente Rotterdam, Rotterdam Climate Initiative, and 100 Resilient Cities, 'Rotterdam Resilience Strategy', pg. 6.

140 Rotterdam Resilience Strategy, pg. 18.

in Rotterdam's climate change strategy. In the process of developing its resilience strategy, the city government began to approach climate mitigation and adaptation as part of a broader attempt to build the city's resilience.[141] As set out in its resilience strategy, Rotterdam's vision is that in 2030, the city will be one where 'the energy infrastructure provides for an efficient and sustainable energy supply' and 'climate adaptation has penetrated into mainstream city operations'.[142] Thinking in terms of resilience has helped the city create synergistic links between climate change, social inclusion, health, and a host of other dimensions that make up the fabric of a city's life.[143]

4.5.1 *Urban Climate Mitigation*

Compared to other Dutch cities like Amsterdam, Rotterdam had a late start in addressing climate change.[144] In 2002, the Klimaatcovenant – a national multilevel arrangement involving local government, provinces, and several ministries – was launched.[145] The Klimaatcovenant provides local authorities additional funding targeted directly at climate change mitigation; these subsidies were instrumental in steering Rotterdam towards developing a climate policy and building capacity to implement specific measures. With national funding, Rotterdam has implemented some twenty projects; these range from awareness raising amongst city officials on the impacts of their commuting habits to the use of residual industrial heat in housing projects.[146]

Tackling climate change rose on Rotterdam's political agenda after representatives of C40 and the Clinton Climate Initiative (CCI) made a compelling case during an official visit to the city in the mid-2000s.[147] They convinced the city's mayor at that time, Ivo Opstelten, that Rotterdam had to prioritize climate adaptation and that the city could make an important contribution towards

[141] Interview No. 11. [142] Rotterdam Resilience Strategy, pg. 24.

[143] See Section 6.3.2.1 of Chapter 6 for more discussion about Rotterdam's resilience strategy.

[144] Joyeeta Gupta, Ralph Lasage, and Tjeerd Stam, 'National Efforts to Enhance Local Climate Policy in the Netherlands' (2007) 4 *Environmental Sciences* 171, pg. 174.

[145] OECD, 'Competitive Cities and Climate Change' (Competitive Cities and Climate Change, Milan, Italy, 9–10 October 2008), pg. 142. Under the Klimaatcovenant framework, subsidies amounting to some 36 million euros were made available for the initial five years. This provided for several hundred local climate assessments and more than 250 municipal implementation plans. The second phase of the scheme was launched in 2008 and included another 35 million euros of subsidies up to 2011; 'Competitive Cities and Climate Change', pg. 147.

[146] Interview No. 11; Gupta, Lasage, and Stam, pg. 175. Most of these climate initiatives also aim to improve Rotterdam's air quality, which is much poorer compared to the rest of the Netherlands because of emissions from industries in the port area and from port-related transport, high population density, and limited amount of green spaces. Population exposure to particulate matter ($PM_{2.5}$) in Rotterdam is 50 per cent higher than in the average OECD port region; see Merk and Notteboom, pg. 11.

[147] Interview No. 11.

climate mitigation by focusing on its port's carbon footprint.[148] In the process, Rotterdam was well placed to share its experience and knowledge with other global port cities. C40 invited Rotterdam to join the network, and this gave the additional impetus that motivated Opstelten's government to act on climate change.[149] Soon thereafter, the Rotterdam Climate Initiative (RCI) was launched in 2006.[150] RCI brings together the Port of Rotterdam Authority, the regional environmental protection agency (DCMR Rijnmond), the city government, and Deltalinqs (an organization that represents the business and industry sector of Rotterdam) to develop and implement Rotterdam's climate change strategy.[151] Mayor Opstelten committed the city to a target of 50 per cent reduction of carbon dioxide emissions in 2025 compared to 1990.[152] The Rotterdam Climate Office, an agency tasked with implementing the RCI, was the first of its kind in the Netherlands.[153]

4.5.2 *Reducing the Carbon Footprint of Europe's Largest Port*

Industry and energy-generating facilities in the port area are responsible for nearly 90 per cent of Rotterdam's carbon dioxide emissions.[154] Rotterdam will not be able to meet its '50% by 2025' target without significant efforts to reduce port-related emissions. In this respect, since late 2000s, the port of Rotterdam has invested significantly in developing its capabilities in energy efficiency, production of renewable energy, and carbon capture and storage (CCS), thereby earning itself an international reputation as a model 'green port'.[155] The port's key energy efficiency initiative lies in the development of GHG-neutral networks and, in particular, a system that transmits heat amongst the firms operating in the port area via a pipeline. In the area of renewable energy, in 2009, the port signed an agreement for the extension of windmill parks that will double the production of wind energy from 151 MW to 300 MW between 2009 and 2020.[156] Further, since 2007, there have been ongoing experiments in using onshore electricity for inland barges. Rotterdam's port is also working with a consortium of private firms to develop technologies and knowledge that would allow for the capture and storage of carbon dioxide under the North Sea.[157] The Rotterdam Capture and Storage Demonstration Project (ROAD) is one of the largest CCS demonstration projects in the world.[158]

[148] Ibid. [149] Ibid.; Gupta, Lasage, and Stam, pg. 176.
[150] Rotterdam Climate Initiative, online: www.rotterdamclimateinitiative.nl (accessed on 28 September 2014).
[151] Ibid.
[152] Rotterdam Climate Initiative, online: www.rotterdamclimateinitiative.nl/en/50procent-reduction (accessed on 28 September 2014).
[153] Mees and Driessen, pg. 272.
[154] Rotterdam Climate Initiative, CO2 Monitor 2014, version 1.0 (10 June 2015), pg. 1.
[155] Ibid., pg. 71. [156] Ibid., pg. 71. [157] Rotterdam, pg. 27.
[158] Ibid.; Rotterdam Capture and Storage Demonstration Project, online: http://road2020.nl/en/ (accessed on 1 October 2015).

While Rotterdam was (and remains) keen to address the climate impact of port and shipping activities, it also had to keep another important consideration in mind – that is, any measures to improve the port's environmental performance should not hinder its competitiveness. As Fenton points out, '[t]his meant cooperation with other ports and stakeholders would be essential, as individual actions by a lone first-mover potentially risked generating negative socio-economic impacts, whilst moving environmental problems elsewhere. Other ports faced similar challenges, making collective action essential and win-win solutions desirable.'[159] Consequently, Rotterdam approached C40 and CCI to ask for support in developing a programme for world ports to address climate change as well as build political support in large port cities for such climate initiatives.[160] This marked the beginning of a partnership between Rotterdam and C40 to develop a World Ports Climate Conference and Declaration, which has since continued fulfilling its agenda as the World Ports Climate Initiative.[161] The conference took place in Rotterdam on 9–11 July 2008. It was well attended by representatives from many of the world's largest ports, shipping companies, terminal operators, fuel suppliers, and environmental NGOs. The conference culminated in the adoption of the World Ports Climate Declaration, whereby ports pledged to use the declaration to 'guide action to combat global climate change and improve air quality'.[162]

4.5.3 *A Leader in Climate Adaptation*

Nearly half of the Netherlands lies below sea level, rendering the country highly vulnerable to sea level rises and other impacts of climate change. Adaptation is therefore a pertinent reality, particularly for delta cities like Rotterdam. In the spirit of innovation and turning risks into opportunities, Rotterdam has developed and implemented numerous adaptation measures that have earned the city both national and global recognition as a leader in climate adaptation.[163]

[159] Paul Fenton, 'The Role of Port Cities and Networks: Reflections on the World Ports Climate Initiative' (Shipping in a Changing Climate – Liverpool, 18–19 June 2014), pg. 7.

[160] Ibid.

[161] The Initiative continues under the auspices of the International Association of Ports and Harbours, which, inter alia, organizes biannual conferences, maintains a dedicated website, disseminates information, and develops projects such as the Environmental Shipping Index (ESI); World Ports Climate Initiative website: http://wpci.iaphworldports.org (accessed on 1 October 2014).

[162] World Ports Climate Declaration, online: http://wpci.iaphworldports.org/data/docs/about-us/Declaration.pdf (accessed on 1 October 2016).

[163] In the first systematic survey of the climate mitigation and adaptation activities of the twenty-five largest municipalities in the Netherlands (population over 100,000), den Exter et al. identify Rotterdam as one of four best-performing cities (alongside Amsterdam, Tilburg, and The Hague). Furthermore, Rotterdam is the only city recognized for its innovative and effective adaptation activities (e.g., flood management, stormwater storage, and adaptive building practices); Renske den Exter, Jennifer Lenhart, and Kristine Kern, 'Governing Climate Change in Dutch Cities: Anchoring Local Climate Strategies in Organisation, Policy and Practical Implementation' (2014) *Local Environment: The International Journal of Justice and Sustainability*, pg. 6.

Rotterdam is one of only three cities in the Netherlands to have a clear adaptation goal, which is to be '100% climate proof in 2025'.[164] The *Rotterdam Climate Change Adaptation Strategy* explains that the term 'climate proof' has a two-pronged meaning: (1) that by 2025, the city would already have taken measures 'to ensure that every specific region is minimally disrupted by, and maximally benefits from, climate change'; and (2) that the city authorities will 'structurally [take] into account the long-term foreseeable climate change in all spatial development of Rotterdam'.[165] In terms of the institutional framework, the 'Rotterdam Climate Proof' programme is part of the Rotterdam Climate Initiative and was established with a budget of 31 million euros for the first four years (2008–2012).[166] That there are six 'Rotterdam Climate Proof' staff dedicated to adaptation planning is rather unusual and symbolizes the political importance attached to the climate programme.[167]

The basis of Rotterdam's adaptation strategy is to build upon its current system for the supply of urban water and flood protection, which consists of storm surge barriers and dikes, canals and lakes, sewers and pumping stations. However, the innovation of Rotterdam's approach lies in the city's attempt to integrate adaptation projects into the urban landscape and improve quality of life for the city's residents. 'Water squares', such as the Bellamyplein water square, serve as both water storage facilities as well as attractive public spaces for residents to enjoy their leisure time. In the Eendragtspolder district, extra water storage has been intelligently combined with a rowing course and other sport facilities.[168] The city has also increased the amount of natural vegetation and flora to combat the urban heat effect – a 'no regrets measure' which has also improved air quality and made Rotterdam a more attractive place in which to live and work.[169] Rotterdam's approach of leveraging more 'green solutions' to protect residential areas rather than implement high-technology and costly solutions has had strong appeal for other delta cities, which are keen to learn from Rotterdam's experience.[170]

Finally, Rotterdam is also the founder of 'Connecting Delta Cities' (CDC), a network within C40's Water and Adaptation Initiative that enables delta cities to share knowledge and best practices in developing their adaptation strategies.[171] Amongst the global cities that are vulnerable to rising sea levels and are CDC

164 City of Rotterdam, *Rotterdam Climate Change Adaptation Strategy* (2013). The other two cities are The Hague and Tilburg; see Renske den Exter, Jennifer Lenhart, and Kristine Kern, pg. 7.

165 City of Rotterdam, ibid., pg. 22.

166 Rotterdam Climate Proof (2009 Adaptation Programme), online: www.rotterdamclimateinitiative.nl/documents/RCP/English/RCP_adaptatie_eng.pdf (accessed on 3 October 2016).

167 Mees and Driessen, pg. 275.

168 City of Rotterdam, pg. 28.

169 Ibid., pg. 29. Also see 'Can Rotterdam Become the World's Most Sustainable Port City?' *CNN* (26 August 2013).

170 Connecting Delta Cities newsletter, 'International Delta Conference in Rotterdam', online: www.deltacities.com/newsletter/international-delta-conference-in-rotterdam?news_id=66 (accessed on 1 October 2016).

171 Connecting Delta Cities, online: www.deltacities.com/about-c40-and-cdc (accessed on 1 October 2016).

members are Jakarta, New Orleans, London, Melbourne, Copenhagen, and Tokyo. CDC links cities at the policy level through bilateral MOUs and letters of intent between the CDC cities. To facilitate the flow of information between CDC cities, a small secretariat has been installed in Rotterdam.

4.6 SEOUL

The largest city and capital of South Korea, Seoul is home to 25 per cent of the country's population.[172] Its 10.4 million residents live within 605 square kilometres, making Seoul one of the most densely populated cities in the world.[173] It is also a wealthy city that has prospered greatly since the 1960s, after the Korean War ended and the country began a process of reconstruction. In this regard, it should be noted that in the 1960s, '[South Korea] was poorer than Bolivia and Mozambique; today, it is richer than New Zealand and Spain, with a per capita income of almost [US]$23,000.'[174] As such, amongst Asian cities, Seoul is in the unique and enviable position of having the financial resources (as well as political leadership and impressive record of technological innovation) to undertake ambitious climate change action. To indicate the extent of Seoul's resources, the city's administrative body, the Seoul Metropolitan Government, had an operating budget of US$18.4 billion in 2010.[175]

However, like many cities that have enjoyed rapid economic growth, Seoul has serious environmental problems such as severe air pollution.[176] Seoul is also highly vulnerable to the impacts of climate change, particularly rising temperatures and the urban heat island effect, increased risks of flash floods, and rainstorms.[177] Research has shown that Seoul has registered a 1.5 °C temperature increase in a period of just thirty years – two-and-a-half times more than the temperature increase of the surrounding rural areas.[178] The daily mortality rates for Seoul as a result of increased heat ranges from 2.7 to 16.3 per cent.[179] Since the mid-2000s, Seoul has attracted

[172] Seoul Metropolitan Government, 'Ranking', online: http://english.seoul.go.kr/gtk/gcs/ranking.php (accessed on 20 October 2016).
[173] Ibid. [174] Marcus Noland, 'Six Markets to Watch: South Korea' [2014] Foreign Affairs.
[175] Seoul Metropolitan Government, 'Ranking', online: http://english.seoul.go.kr/gtk/gcs/ranking.php (accessed on 20 October 2016).
[176] In addition to high volumes of traffic emissions, which is a common cause of poor air quality in highly congested cities, Seoul is badly affected by the dust clouds from Mongolia and China that arrive annually and engulf the city in a fine yellow dust; see, for example, Alex Kirby, 'Asia's Dust Storm Misery Mounts', BBC News, 31 March 2004, online: http://news.bbc.co.uk/2/hi/science/nature/3585223.stm (accessed on 20 October 2016).
[177] H. Lee and J. G. Oh, 'Integrating Climate Change Policy with a Green Growth Strategy: The Case of Korea' in Bryce Wakefield (ed), *Green Tigers: The Politics and Policy of Climate Change in Northeast Asian Democracies* [Woodrow Wilson International Center for Scholars (Asia Program Special Report) 2010], pg. 18.
[178] Barry Munslow and Tim O'Dempsey, 'Globalisation and Climate Change in Asia: The Urban Health Impact' (2010) 31 *Third World Quarterly* 1339, pg. 1343.
[179] Ibid.

global attention for its 'low-carbon green growth' policies and programmes as the city commits itself to reducing its significant GHG emissions and adapting to climate change. In 2009, Seoul hosted the third Summit of the C40 Large Cities Climate Leadership Group, an event that 'gave visibility to Seoul's and Korea's climate change initiatives while at the same time applying pressure on the city and the nation to perform'.[180] While Seoul's commitment to addressing climate change is very much influenced by national policies and laws, its involvement in transnational climate change networks has provided valuable access to information about how other cities were performing and to learn from their experiences. In the words of Deputy Mayor of Seoul Kim Sang Bum during an interview at the C40 Mayors Summit in Johannesburg (4–6 February 2014), '[w]hen Seoul hosted the C40 Mayors Summit in 2009, we had little to no experience with how to improve energy efficiency in buildings'; it was through the process of sharing knowledge with other C40 cities that Seoul gained the confidence and know-how to improve its energy efficiency.[181]

The following discussion proceeds by first briefly setting out the national context for Seoul's climate change initiatives, because Seoul has been the de facto national laboratory for the country's climate change policies and programmes. Korea's climate change strategy places a heavy emphasis on technological advancement, fostering research and development, and capital investment in climate-proof infrastructure. A significant proportion of resources has inevitably been channelled to Seoul, the country's most advanced and populous city. Thus, an understanding of the national climate policy would shed light on Seoul's approach. At the same time, the city's metropolitan government and mayor actively participate in numerous pan-Asian and transnational city climate change networks which exercise a strong motivational influence on Seoul's quest to reinvent itself as a 'global climate-friendly city' (a term borrowed from its *Low Carbon Green Growth Master Plan*). The section then describes Seoul's mitigation and adaptation programmes, some of which have garnered international awards and recognition.

4.6.1 *Korea's Low-Carbon Green-Growth Vision*

As a non-Annex I party to the Kyoto Protocol, Korea did not face any international legal obligations to reduce its GHG emissions. However, since the negotiations of

[180] Miranda A. Schreurs, 'Multi-Level Governance and Global Climate Change in East Asia' 5 *Asian Economic Policy Review* 88, pg. 100. Acuto argues that the hosting of key C40 events by Tokyo (in 2008), Seoul (in 2009), and Hong Kong (in 2010) is an indication of how key global cities in Asia progressively gained prominence in the C40 network, possibly influencing the network's pace and agenda; Michele Acuto, 'The New Climate Leaders?' (2013) 39 *Review of International Studies* 835, pg. 853.

[181] C40 Summit Video Blog Series: Kim Sang Bum, Deputy Mayor of Seoul – 'Sharing Knowledge with Other Cities Is One of the Benefits of C40', online: www.c40.org/blog_posts/c40-summit-video-blog-series-kim-sang-bum-deputy-mayor-of-seoul-sharing-knowledge-with-other-cities-is-one-of-the-benefits-of-c40 (accessed on 20 October 2016).

the Kyoto Protocol, the Korean economy has grown and the country now has a higher per capita GDP than approximately half of the Annex I parties.[182] It is also the world's fifteenth-largest source of emissions.[183] In the negotiations for the post-2012 climate change legal regime, it was assumed that Korea will be regarded as a developed country and expected to undertake binding GHG reduction commitments just like its developed-country counterparts.[184] In this context, under the Paris Agreement, Korea has pledged to reduce its GHG emissions by 37 per cent of 'business as usual' estimates by 2030.[185]

Korea embarked on a national 'green growth' strategy as a response to climate change and the 2008 global financial crisis. In August 2008, President Lee Myung-bak launched the Low Carbon Green Growth vision. Initially formulated as a US$38.5 billion 'Green New Deal' to lift the Korean economy out of recession, the Low Carbon Green Growth strategy was subsequently reframed as the nation's vision to guide development for the next fifty years.[186] This vision has since been implemented via the National Strategy for Green Growth, the Five Year Green Growth Plan (2009–2013), and an array of institutional programmes to ensure inter-ministerial cooperation and effective implementation at the local level.[187]

The green-growth strategy also aims to reduce Korean dependence on energy imports, which account for 86 per cent of the country's primary energy supply, and to improve quality of life for Koreans by reversing some of the worst environmental degradation.[188] However, the emphasis is on green technologies as new engines of growth and job creation for an economy facing rapid population ageing and slowing growth. As Matthews elaborates,

> [w]hat is distinctive about Korea's approach to [green growth] is that it is an industrial strategy ... framed around the promotion of key technologies and industries that are viewed as providing the growth engines for the next stage of Korea's development, and as export platforms for the 21st century, as well as means

182 Central Intelligence Agency (US), 'The World Factbook', online: www.cia.gov/library/publications/the-world-factbook/geos/ks.html (accessed on 20 October 2016).

183 R. S. Jones and B. Yoo, *Achieving the 'Low Carbon, Green Growth' Vision in Korea* (OECD Economics Department Working Paper No. 964, 2012), pg. 8.

184 J. H. Lee, John M. Leitner, and Minjung Chung, 'The Road to Doha through Seoul: The Diplomatic and Legal Implications of the Pre-COP 18 Ministerial Meeting' (2012) 12 *Journal of Korean Law* 55, pg. 66.

185 Submission by the Republic of Korea, 'Intended Nationally Determined Contribution', online: www4.unfccc.int/Submissions/INDC/Published%2520Documents/Republic%2520of%2520Korea/1/INDC%2520Submission%2520by%2520the%2520Republic%2520of%2520Korea%2520on%2520June%252030.pdf (accessed on 2 November 2016).

186 Jones and Yoo, pg. 6. 187 Ibid., pg. 5.

188 Ibid. The Four Major Rivers Restoration Project, hailed as a major climate change adaptation project to mitigate flood risks and to restore water quality, has been controversial because of its ambiguous environmental benefits: see James Card, 'Korea's Four Rivers Project: Economic Boost or Boondoggle?' *Yale Environment* 360, 21 September 2009, online: http://e360.yale.edu/feature/koreas_four_rivers_project_economic_boost_or_boondoggle/2188/ (accessed on 22 October 2016).

to reduce carbon emissions. Although climate objectives are mentioned prominently, it is fundamentally an industrial upgrading strategy.[189]

Since 2008, there has been a sharp increase in investments by Korean firms in 'green industries'. The major investment areas have been renewable energy, next-generation electric equipment, and green cars. According to a government survey, investment by the thirty largest businesses was three times higher in 2010 than in 2008. This amounted to US$13.7 billion over the three-year period (1.5 per cent of 2009 GDP). Korea is now the world's second-largest producer of lithium rechargeable batteries and LED devices. Economists are of the view that it is still too early to assess the impact of the Green Growth Strategy on economic growth.[190] However, for current purposes, the salient point is that Seoul benefits from a strong national political and business consensus on the value of green growth and tackling climate change, rendering it easier to foster private-public, multi-stakeholder partnerships to implement its climate change programmes.

4.6.2 'A *Global Climate-Friendly City by 2030*'

Seoul was the first city in Korea to establish a climate mitigation and adaptation strategy, which is known as the 'Master Plan for Low Carbon Green Growth'. The key features of Seoul's Low Carbon Green Growth Master Plan are as follows:[191]

- By 2030, the city aims to reduce GHG emissions by 40 per cent from 1990 levels, reduce energy consumption by 20 per cent, and increase renewable energy use by 20 per cent.
- By 2030, all new buildings will be required by law to attain green building certification.
- By 2030, buildings older than 20 years will be required to undergo mandatory energy auditing.
- By 2030, all public transportation will rely on electric and other green vehicles.
- By 2020, the aim is to increase bike ridership to 10 per cent and public transport usage to 70 per cent.
- By 2020, all street lighting and indoor lighting in public buildings will use LED.
- Seoul will revise its urban planning laws to promote the addition of eleven square kilometres of green space to the city.

[189] John A. Mathews, 'Green Growth Strategies – Korean Initiatives' 44 *Futures* 761, pg. 762.
[190] Kang, Jin-gyu, and Hongseok, pg. 13.
[191] The information that follows is reproduced from the Master Plan, a copy of which can be found online: http://planning.cityenergy.org.za/Pdf_files/world_cities/asia/city_of_seoul/Seoul%2520Climate%2520Change%2520Action%2520Plan.pdf (accessed on 22 October 2016).

- By 2030, Seoul aims to invest US$2 billion in research and development (R&D) of selected green technologies and support eventual commercialization. These technologies are next-generation hydrogen fuel cell, photovoltaic cell, 'smart grid' technologies (the use of information technologies in buildings to maximize energy efficiency), LED lighting, electric cars, urban environmental restoration systems, waste-to-energy technologies, and climate change adaptation technologies. In addition, the metropolitan government will establish partnerships and platforms for business, academia, and other stakeholders to facilitate R&D. These partnerships and platforms extend from the city level to transnational research consortiums such as the Global CCS Institute.[192]

The Master Plan also identifies five focal areas for adaptation: infectious diseases (certain disease infection rates will increase because of the warmer climate), rising urban temperatures and heat waves, significant fluctuations in water supply ranging from drought to flooding, and disruptions to ecosystems.

Within a short span of five years since the announcement of the Master Plan in 2009, an impressive number of high-profile climate projects have been successfully implemented. Three projects that have received international attention because they showcase best practices and feasibility are briefly described further on. These examples have been shared at international conferences, workshops, publications, and databases (such as Connected Urban Development, C40, and ICLEI – Local Governments for Sustainability, of which Seoul is a member).[193]

Hydrogen Fuel Cells: Seoul has been increasing its capabilities to generate more power from hydrogen fuel cells. The city aims to meet 10 per cent of its total energy needs from hydrogen fuel cells by 2030 – enough power to supply some 400,000 households. By 2014, Seoul has opened 29 fuel cell power stations. It is supporting the deployment of fuel cells in commercial buildings with a combination of subsidies, technical support, and low-interest loans. This project qualified as a finalist in the green energy category of the 2013 City Climate Leadership Awards, which confer global recognition on cities that are demonstrating leadership in addressing climate change.[194]

[192] The Global CCS Institute's mission is to accelerate the development, demonstration, and deployment of carbon capture and storage. It is headquartered in Melbourne, Australia, where it was established in 2009 with initial funding from the Australian government; online: www.globalccsinstitute.com (accessed on 22 October 2016).

[193] Seoul is a founding city of Connected Urban Development (CUD). CUD brings cities, business partners, and NGOs together in a global platform committed to the use of information and communications technology in urban infrastructure to reduce carbon emissions (online: www.connectedurbandevelopment.org).

[194] This paragraph is derived from the information found on the 2013 City Climate Leadership Awards finalists webpage, online: http://cityclimateleadershipawards.com/seoul-hydrogen-fuel-cells/ (accessed on 22 October 2016).

Car-Free Days: This programme has effectively reduced Seoul's annual carbon dioxide emissions by 10 per cent – about 2 million tonnes of carbon dioxide every year – by keeping 2 million cars off the road. Drivers can select one day a week as their 'no driving day' on a website (www.no-driving.seoul.go.kr) and receive an electronic tag. Once registered in this programme, public-sector participants qualify for incentives such as a 5 per cent discount on the annual vehicle tax, a 50 per cent discount on charges levied by the city's congestion charging scheme, and free parking. Private participants qualify for discounts on fuel and car maintenance costs, as well as free car washes. The city monitors compliance through radio frequency identification that detects the electronic tags on participating vehicles.[195]

Star City Rainwater Project: Star City is a major real estate development project with more than 1300 apartment units in Gwangjin-gu, a district in Seoul. In 2007, a rainwater harvesting system with a catchment area of 6,200 square meters of rooftop and 45,000 square meters of terrace was installed. The system collects up to 100 millimetres of rainwater, which is used for gardening and public toilets. The Star City rainwater harvesting system has proven to be a highly effective adaptation response to the risks of flooding, and in the long run, will reduce energy consumption for water treatment and conveyance. The success of Star City motivated the city's metropolitan government to pass a regulation in 2004 that requires all new public buildings, new town projects, and large private buildings to install rainwater harvesting systems. The central disaster prevention agency located in the city government headquarters monitors the water levels in all water tanks and, if required, issues orders to building owners to empty their rainwater tanks in anticipation of heavy rainfall.[196]

4.7 CONCLUSION

In this chapter, we learned about the policies, strategies, and programmes that five global cities – London, Mexico City, New York City, Rotterdam, and Seoul – have put in place in response to the risks of climate change as well as to reduce their GHG emissions. As much of the chapter pays attention to local circumstances and efforts, it is easy to get lost in the details and lose sight of how this chapter fits into the larger narrative of this book. Thus, this conclusion will highlight a few salient points that arise from the preceding discussion and relate them to the central

[195] This paragraph is derived from the information found on the C40 case study section, 'Seoul Car-Free Days Have Reduced CO2 Emissions by 10% Annually', online: http://c40.org/case_studies/seoul-car-free-days-have-reduced-co2-emissions-by-10-annually (accessed on 22 October 2016).

[196] This paragraph relies on the information found in M. Y. Han, J. S. Mun, and H. J. Kim, 'An Example of Climate Change Adaptation by Rainwater Management at the Star City Rainwater Project' (3rd IWA Rainwater Harvesting Management International Conference, Gyeongnam, Goseong County, Republic of Korea, 20–24 May 2012), online: www.iwahq.org/ContentSuite/upload/iwa/Document/session%2520a%252001.pdf (accessed on 22 October 2016).

analysis of the emergence of global cities as governance actors who are beginning to exercising lawmaking functions in the transnational climate change regime complex.

First, the purpose of this chapter has been to inform the reader of the extent and type of governance activities global cities around the world are engaging in. From the outset, Mayor Livingstone of London recognized that cities ought to learn from one another, inspire and support each other's efforts and, in the process of repeated interactions, cultivate norms and practices concerning urban climate mitigation and adaptation. That cities can play a meaningful role in addressing a global environmental problem like climate change even when states appear incapable of taking concerted action is a norm that began to emerge during Livingstone's mayoralty. It soon became internalized by city officials and other actors and gained transnational traction especially as a result of C40's public relations campaigns, as we will see in Chapter 5. Eventually, it has become a norm within the transnational climate change regime complex. In this regard, the local has influenced the global, just as the global agenda has shaped the local one.

Secondly, a factor that has enabled cities to play a role in governing climate change is the resources made available by transnational actors, including international organizations, private foundations, and global environmental NGOs. For example, Mexico City's Metrobus System project, which has significantly improved air quality and reduced GHG emissions in the city, enjoyed the support of World Resources Institute (a leading global environmental NGO) and funding from the World Bank, the Shell Foundation, and the Caterpillar Foundation.[197] The availability of financial support from external parties will, of course, be particularly relevant for less-wealthy cities. However, even for wealthier cities such as Rotterdam in the Netherlands, support from transnational actors has also made a difference. For example, Rotterdam approached C40 to initiate a programme for ports to address climate change because it did not want to go at it alone at the risk of its port losing its competitive advantage. The city's involvement in the Rockefeller Foundation's 100 Resilient Cities programme also significantly shaped and has had lasting influence on Rotterdam's strategic approach and programmes on climate mitigation and adaptation. In the case of Seoul, the city had the financial resources but lacked the policy and technical know-how. Participation in a transnational network, C40, helped Seoul gain critical knowledge and overcome its learning curve more quickly than it would have otherwise.[198] By lending financial and technical assistance to cities, global civil society actors, private foundations, and international organizations help shape the transnational climate governance agenda by spreading policy concepts such as resilience. They also exercise considerable influence on a city's low-carbon devel-

[197] See footnote 85 of this chapter. [198] Interview No. 1, on file with author.

opment choices when they decide on the projects that will receive funding. As such, technical and financial assistance is not neutral and free from ideological influences and politics.

Finally, this chapter captures some of the innovative experimentation that is currently happening in cities across the world. These experiments contribute towards transnational climate change governance when global cities seek to scale up their actions, pursue cooperation, and develop harmonized standards through cross-border networks. In Chapter 5, we turn our attention to one such network that has become highly influential in the transnational climate change governance landscape within a short span of time: C40.

5

Transnational Urban Climate Governance via Networks: The Case of C40

5.1 INTRODUCTION

In their attempts to reduce GHG emissions and adapt to the impacts of climate change, cities have found it helpful to establish networks through which they can facilitate policy learning and develop new governance approaches.[1] The information and communications revolution has made such networking easier, more affordable, and quicker. These networks connect city officials across the world; they are therefore cross-border and transnational in nature.[2] Transnational municipal networks also serve as conduits through which cities can create and implement urban-specific norms, practices, and voluntary standards that support and complement the international legal regime on climate change. In addition, these networks have formed linkages and partnerships with national authorities, international organizations, multinational corporations, and civil society. They therefore also perform the function of linking cities to other actors in the transnational climate change regime complex. By disseminating knowledge about urban climate change practices to other governance actors through these networks, cities have the potential to shape the norms and practices of other actors.[3]

This book is not concerned with cities broadly speaking, but rather with global cities that command significant economic and political resources and are leading

[1] Bouteligier points out that such networks allow for the conceptualization of cities, traditionally linked to local policies, as actors in global governance; Sofie Bouteligier, 'Inequality in New Global Governance Arrangements: The North-South Divide in Transnational Municipal Networks' (2013) 26 *Innovation: The European Journal of Social Science Research* 251, pg. 252.

[2] That cities are forming networks to address climate change is consistent with the trend (that emerged in the 1990s) of cities actively forming networks to address common environmental problems such as air pollution and biodiversity loss. This trend is mostly ascribed to chapter 28 of Agenda 21, which recognizes the role of local authorities in the promotion of sustainable development and advocates exchange and cooperation between them; Agenda 21: Programme of Action for Sustainable Development, U.N. GAOR, 46th Sess., Agenda Item 21, UN Doc A/Conf.151/26 (1992), online: https://sustainabledevelopment.un.org/content/documents/Agenda21.pdf (accessed on 10 July 2016).

[3] Toly notes that the horizontal and vertical relationships fostered by a transnational city network 'may permit cities a significant role in the diffusion of both techniques and norms, important functions of governance and politics'; Noah J. Toly, 'Transnational Municipal Networks in Climate Politics: From Global Governance to Global Politics' (2008) 5 *Globalizations* 341, pg. 344.

the current wave of urban climate action. This chapter therefore focuses on C40, which has become widely recognized, within a decade of its founding in 2005, as the leading network of global cities addressing climate change.[4] C40 positions itself as a gathering of the top echelon of the world's major cities, a 'space of engagement' for cities gathered to exchange expertise and knowledge on climate change as well as a catalyst representing (and creating more) connections amongst major cities.[5] In seeking to present cities as significant actors in the global response to climate change, C40 has emphasized the global nature of its cities, because global cities are widely understood to be of pivotal importance in the global economy and enjoy a higher status compared to other cities.[6] C40 seeks to support the UNFCCC regime but also underpins its legitimacy on the claim that cities are forging ahead with practical and innovative climate mitigation and adaptation measures while international climate negotiations have proceeded at a glacial pace (at least until the COP in Paris in December 2015).[7] One of C40's central messages is that 'cities act while states only talk'. As former chairman of C40 and current UN Special Envoy on Cities and Climate Change, Michael Bloomberg, has put it, 'Cities are where you deliver services. Federal governments and state governments sit around talking and passing laws or recommendations that don't have any teeth.'[8]

C40 seeks to play the role of a knowledge broker in the transnational climate change governance arena. C40's aspiration is that, through its role as an information-sharing platform, individual global cities will be empowered with technical knowledge, shared resources, and technology that the cities would not be able to obtain on

[4] C40 also enjoys significant coverage by highly respected global media outlets; see, for example, 'Greening the Concrete Jungle' *The Economist* (3 September 2011), online: www.economist.com/node/21528272; Rene Vollgraaff and Janice Kew, 'Cities Almost Double Climate Actions over 2 Years, C40 Says' *Bloomberg* (5 February 2014), online: www.bloomberg.com/news/2014-02-05/cities-almost-double-climate-actions-over-two-years-c40-says.html; Alison Kemper and Roger Martin, 'Cities Are Businesses' Best Allies in the Battle against Climate Change' *The Guardian* (14 October 2014), online: www.theguardian.com/sustainable-business/2014/oct/14/cities-businesses-best-allies-battle-against-climate-change (all three articles accessed on 23 December 2014). Further, C40 has been put forward as part of a 'Coalition of the Working between countries, companies and cities', to inject energy into the repeatedly stalled multilateral efforts and to prompt practical action, in the highly influential and authoritative report of the Oxford Martin Commission for Future Generations, 'Now for the Long Term' (Oxford Martin School, October 2013), online: www.oxfordmartin.ox.ac.uk/downloads/commission/Oxford_Martin_Now_for_the_Long_Term.pdf (accessed on 30 November 2016).

[5] The term 'space of engagement' is from Kevin R. Cox, 'Spaces of Dependence, Spaces of Engagement and the Politics of Scale, or: Looking for Local Politics' (1998) 17 *Political Geography* 1.

[6] Michele Acuto, 'The New Climate Leaders?' (2013) 39 *Review of International Studies* 835, pg. 841. Lee has sought to quantitatively test the hypothesis that global cities are more likely to become members of transnational climate change networks because, in addition to their economic interests regarding climate change, 'they provide underlying conditions facilitating information sharing and diffusion' as hubs of international economic and policy interactions; Taedong Lee, 'Global Cities and Transnational Climate Change Networks' (2013) 13 *Global Environmental Politics* 108, pg. 109.

[7] Sofie Bouteligier, *Cities, Networks and Global Environmental Governance: Spaces of Innovation, Places of Leadership* (Routledge 2013), pg. 85.

[8] Quoted in Alexei Barrionuevo, 'World Bank to Help Cities Control Climate Change' *New York Times* (1 June 2011).

their own. Through such empowerment, global cities will be better placed to engage in policy experimentation and to implement transformative climate mitigation and adaptation projects. Another key aspect of C40's governance mode is to facilitate corporation between its member cities and other transnational actors. This includes establishing and coordinating public-private partnerships with multinational corporations like Siemens, NGOs such as the World Resource Institute, and international financial institutions like the World Bank.[9] In early 2016, C40 formed a partnership with the International Cleantech Network – a transnational network of clean technology cluster organizations – to create the City Solutions Platform. The Platform serves as a forum of engagement for clean technology developers, entrepreneurs, and city authorities seeking clean technology solutions for climate mitigation and adaptation.[10] In addition to increasing C40's capacity to assist its members in making GHG emission reductions, these public-private partnerships (PPPs) enhance the status of C40 as a network that is actively engaged with major political and business actors on the international stage.

From a different perspective, through these PPPs, international organizations are able to enlist C40 as an intermediary in addressing cities in pursuit of the goal of tackling climate change. The C40-World Bank partnership is an interesting example of 'orchestration' whereby the World Bank seeks to initiate, steer, and strengthen governance within the transnational climate change regime complex by, amongst other things, facilitating and endorsing C40's governance efforts. In Chapter 3, there was discussion about the role of the World Bank in creating and sustaining an ideological platform that promotes the linkages between cities, globalization, and development. The World Bank's vision of sustainable urban development is that of 'competitive, well governed, and bankable' cities. In pursuit of this vision, the World Bank has introduced a range of initiatives to help city governments gain access to international financial markets. The range of initiatives to improve the creditworthiness of cities so that they are able to independently raise money to provide services and invest in infrastructure has recently been extended to include the climate change agenda. As will be discussed later on, the C40-World Bank partnership involves giving C40 member cities preferential access to the Low Carbon, Livable Cities initiative, which provides training programmes and technical assistance to improve cities' financial transparency and planning capabilities so that they are better able to obtain financing from public-sector investors and the private sector.

Finally, in the run-up to the UNFCCC COP in Paris, C40 was actively involved in various partnerships to promote the development of standardized GHG accounting methodologies for cities. This will be discussed briefly in this chapter to lay the

[9] C40, 'Our Partners and Funders', online: www.c40.org/partners (accessed on 23 October 2016).

[10] Interview No. 8; International Cleantech Network, 'Collaboration between C40 and ICN: When Cities Meet Clean Technologies' (Copenhagen, 5 December 2015), online: http://internationalcleantechnetwork.com/project/collaboration-between-c40-and-icn-when-cities-meet-clean-technologies/ (accessed on 3 July 2016).

foundation for subsequent analysis in Chapter 6, which argues that global cities, through a network like C40, are playing a jurisgenerative function at the transnational level. Section 5.2 of this chapter provides a brief literature review of transnational networks of cities in the area of climate change. Section 5.3 delves into a detailed discussion of C40, including its partnership with the World Bank and its attempts to develop harmonized standards for urban GHG accounting. Section 5.4 draws concluding remarks about the role of C40 as a network through which global cities exercise governance functions.

5.2 A BRIEF INTRODUCTION TO TRANSNATIONAL MUNICIPAL NETWORKS IN THE AREA OF CLIMATE CHANGE

According to Kern and Bulkeley, transnational municipal networks have three defining characteristics. First, member cities are free to join or leave the network.[11] Secondly, such networks are often characterized as a form of self-governance, as they appear to be non-hierarchical, horizontal forms of governance.[12] Thirdly, decisions taken within these networks are usually directly implemented by member cities.[13]

Cities join transnational networks because they perceive certain advantages such as learning from other cities, expanding their links to international institutions, and gaining preferential access to funding.[14] In large networks, it is not uncommon for there to be a 'core group' of active member cities while the majority of the member cities are relatively passive. In the latter case, membership may be symbolic; for example, a city may have joined the network only because neighbouring cities have done so.[15] Passive members also often lack the financial and human resources required to participate in network activities such as conferences, funding bids, and the implementation of work programmes and standards developed by the network.[16] In contrast, some network members are active locally and transnationally.[17] They take proactive steps to develop the network by, for example, organizing and hosting workshops and frequently uploading best practice case studies on the network's

[11] Kristine Kern and Harriet Bulkeley, 'Cities, Europeanization and Multi-Level Governance: Governing Climate Change through Transnational Municipal Networks' (2009) 47 *Journal of Common Market Studies* 309, pg. 309.

[12] Ibid. [13] Ibid. [14] Ibid.; Interviews Nos. 2 and 3.

[15] In Krause's empirical study of two city networks in the United States, ICLEI and the US Conference of Mayors' Climate Protection Agreement (MCPA), she finds that ICLEI appears to play a more effective role in promoting GHG reductions compared to MCPA. Amongst the reasons is that ICLEI requires cities to achieve certain milestones or targets and to pay an annual membership fee, unlike MCPA, which offers free membership and there is no monitoring of cities' activities. She suggests that weakly committed cities that are motivated by public 'credit claiming' find it easy to join networks with minimal political and financial costs such as MCPA. In such cases, membership in a network does not effectively steer the city towards climate protection; Rachel M. Krause, 'An Assessment of the Impact That Participation in Local Climate Networks Has on Cities' Implementation of Climate, Energy, and Transportation Policies' (2012) 29 *Review of Policy Research* 585, pgs. 587, 601.

[16] Kern and Bulkeley, pg. 327. [17] Ibid., pg. 326.

Internet database. Kern and Bulkeley's findings on transnational city networks in Europe suggest that 'networks are networks of pioneers for pioneers' (pioneers referring to active member cities), and within a network, 'it is easy to distinguish between a hard core of pioneers and a periphery consisting of relatively passive cities which have scarcely changed their behavior since joining the network'.[18]

While scholars like Sassen hypothesize that networks that include cities from the Global North as well as the Global South go beyond the traditional North-South divide, Bouteligier's empirical research on C40 and Metropolis (a global network that covers environmental, social, economic, and cultural issues) demonstrates otherwise.[19] Briefly, the greater involvement of some members (which are usually cities from the Global North that have the human and financial resources to support active involvement) than others concentrates power, 'which results in power relations and the persistence of (structural) inequalities. These inequalities run along dividing lines we know from the past.'[20] Bouteligier finds that those cities that are important to the network in question shape the agenda, determine choices regarding best practices, and influence how problems and solutions are framed. She concludes that, as long as transnational municipal networks perpetuate inequalities, they will face difficulties in dealing with contemporary urbanization problems and therefore will ultimately fail to meet their goals.[21]

Finally, in the early stage of urban climate governance in the 1990s, networks like ICLEI focused on supporting cities in implementing local climate policies. ICLEI's Cities for Climate Protection Program (CCP), established in 1993, is widely recognized to be the first global programme supporting cities in pursuing climate action.[22] CCP's founding document states that one of CCP's key objectives was to enlist 100 municipalities worldwide that emit 1 billion tonnes of global CO_2 (5–10 per cent of the global total) by 1995.[23] To join CCP, a city had to adopt a 'local declaration' committing itself to reducing GHG emissions by meeting five milestones: (1) conducting a baseline energy and emissions inventory, (2) adopting an emissions reduction target, (3) developing a local action plan, (4) implementing policies and measures, and (5) monitoring and verifying outcomes.[24] Betsill and Bulkeley argue

[18] Ibid., pg. 329. [19] Ibid.

[20] Bouteligier, 'Inequality in New Global Governance Arrangements: The North-South Divide in Transnational Municipal Networks', pg. 263.

[21] Ibid., pg. 264.

[22] ICLEI, 'Connecting Leaders: Connect. Innovate. Accelerate. Solve'. online: http://incheon2010.iclei.org/fileadmin/templates/incheon2010/Download/Media_and_Press_Downloads/Connecting_Leaders-www.pdf (accessed on 3 December 2016).

[23] ICLEI, 'Cities for Climate Protection: An International Campaign to Reduce Urban Emissions of Greenhouse Gases', 15 February 1993, pg. 2, online: www.iclei.org/fileadmin/user_upload/ICLEI_WS/Documents/advocacy/Bonn_2014/ADP2.5-Support_Files/ICLEI_TheBirthofCCP_1993.pdf (accessed on 3 December 2016).

[24] ICLEI, *Municipal Leader's Declaration on Climate Change and the Urban Environment*, Article 2 (This declaration is found as an appendix to 'Cities for Climate Protection: An International Campaign to Reduce Urban Emissions of Greenhouse Gases', ibid.).

that '[b]oth the milestone framework and the use of quantification reflect the CCP programme's emphasis on the need to evaluate performance and improve local accountability'.[25] Further, 'through adhering to the CCP program framework and participating in its activities, members share both normative goals, that climate change is a problem and can be addressed locally, and a commitment to a particular policy approach based on the measurement and monitoring of greenhouse gas emissions'.[26] Other networks, including C40, have subsequently embraced this approach. Gordon points out that younger networks like C40 seek to surpass the limitations of the older networks by '[shifting] the emphasis towards coordinating and scaling city climate policy with the aim of achieving aggregate effect, encouraging the diffusion of ideas and practices, and offering an alternative (albeit intersecting) architecture of global climate governance'.[27]

This summary of some of the key findings about how transnational city networks perform in reality provides an introductory platform for the discussion about C40 to follow. Like other transnational city networks, C40 is concerned about laggard cities that do not contribute proactively to achieving the network's mission. As a voluntary tool of governance, transnational city networks depend heavily on persuasion, mutual benefit, and reciprocity to encourage the spread of norms, practices, and voluntary standards amongst member cities. As will be seen in the discussion that follows, C40 seeks to reduce disparities in the commitment and performance of its member cities by being very selective in its membership and providing additional support to less active cities.[28] Transparency mechanisms are also viewed as a means for encouraging passive cities into action, as these reporting platforms provide various stakeholders, including citizens and civil society groups, the means to monitor and hold their cities accountable.[29] In addition, as to one of the key issues raised previously – the North-South divide – it is not surprising that the socio-economic conditions in Global South cities restrict their capabilities of participating as actively as their Global North counterparts in a transnational city network. The salient response is not to lament the potential marginalization of the Global South cities, but to identify ways to create inclusion and equitable participation. First, the principle of common but differentiated responsibilities and respective capabilities (CBDRRC) is a foundational pillar of the international climate change regime and is regarded as key to ensuring equitable and fair treatment. By taking the CBDRRC principle into account in their norm-setting actions, C40 recognizes the need for more developed and wealthier member cities to assist the efforts of poorer and developing cities.[30] Cities with more resources are expected to take the lead in

[25] Michele M. Betsill and Harriet Bulkeley, 'Transnational Networks and Global Environmental Governance: The Cities for Climate Protection Program' 48 *International Studies Quarterly* 471, pg. 478.

[26] Ibid.

[27] David J. Gordon, 'Between Local Innovation and Global Impact: Cities, Networks, and the Governance of Climate Change' (2013) 19 *Canadian Foreign Policy Journal* 288, pg. 293.

[28] See discussion in Section 5.3.3. [29] Ibid. [30] See discussion in Section 6.4.2 of Chapter 6.

organizing and hosting C40 events, and some cities go further by subsidizing the travel and accommodation arrangements of participants from less wealthy cities.[31] Secondly, mayors serve on a rotating basis on C40's Steering Committee to provide overall strategic direction.[32] The mayors on the committee represent various geographical regions to ensure fair representation of all C40 cities. Finally, it has been noted that, in order for networks like C40 to succeed, the interactions of member cities must be pursued in a spirit of mutual respect, appreciation of the differences amongst global cities, and a shared common goal to address climate change.[33] It can be argued that articulating these values as guiding principles will go some way towards creating a firmer normative basis for overcoming some of the North-South inequalities.

5.3 C40

This part of the chapter explores the origins of C40, followed by a brief description of its membership and governance structure. The discussion then turns its focus to C40's climate programmes and initiatives to shed light on the ways in which C40 seeks to fulfil its role of facilitating collaboration amongst global cities and how these cities, in turn, use C40 to exercise agency and influence in the transnational climate change regime complex. The partnership between C40 and the World Bank, as well as C40's attempts at developing a GHG accounting standard, will also be examined here.

5.3.1 *The Origins of C40*

On 3–5 October 2005, at the initiative of then mayor of London Ken Livingstone, the Greater London Authority convened a two-day World Cities Leadership and Climate Summit.[34] This meeting of eighteen major cities was carefully timed to coincide with the Group of Eight (G8) Summit in Gleaneagles, Scotland, the annual gathering of the leaders of the world's major powers to address major economic and political issues.[35] 'Livingstone's original idea was much in parallel with that year's G8 . . . as the Group gathered the largest economies, he gathered the largest cities, where "large" was not just a measure of size but of importance.'[36] Thus, Livingstone's vision was one of 'the elite of core cities' that, because of sheer population and global prominence, would lead the urban response to climate

[31] Interview No. 1. [32] C40, online: www.c40.org/about (accessed on 1 November 2016).

[33] Interview No. 1.

[34] The summit was convened in partnership with ICLEI and the Climate Group, an international non-profit group founded in 2004 that works with states, cities, and businesses to develop climate finance mechanisms and low-carbon business models; online: www.theclimategroup.org (accessed on 27 November 2016).

[35] For analysis of the role of the G8 in international relations, see *New Directions in Global Political Governance: The G8 and International Order in the Twenty-First Century* (John Kirton and Junichi Takase, eds, Ashgate 2002).

[36] Greater London Authority political officer, quoted in Acuto, pg. 854.

change.[37] Originally known as the C20, the cities that took part in the network's inaugural summit included Barcelona, Beijing, Berlin, Brussels, Chicago, London, Madrid, Mexico City, New Delhi, New York, Paris, Philadelphia, Rome, San Francisco, São Paulo, Shanghai, Stockholm, Toronto, and Zurich.[38]

The C20 Climate Change Summit Communique that was issued at the close of the World Cities Leadership and Climate Summit gives some insight into how the founders envisioned the role of the network in the global response to climate change as well as its goals in the short and long term. First, the declaration refers to 'the C20 Large World Cities'.[39] 'The C20 cities recognize [their] role as large city governments' and that C20 will work with ICLEI, which represents 'cities of all sizes'.[40] In the communiqué, the size of the populations that the C20 cities represent is the main basis for differentiating C20 cities from other cities and sub-national governments and is the precursor to subsequent attempts to highlight the global standing of C40 cities in the network's branding strategy.

Secondly, in the communiqué, C20 portrays cities as indispensible partners in the global effort to tackle climate change because cities are both a significant cause and a solution to the problem of climate change. The communiqué states that cities are a significant source of GHGs 'from cars, trucks, industries, manufacturing, buildings and waste' and are 'growing in significance as more of the world's population reside there'.[41] At the same time, '[l]arge cities have sizeable economies that are ideal markets to incubate, develop, and commercialise greenhouse gas reducing and adaptation technologies, including those to improve energy efficiency, waste management, water conservation, and renewable energy'.[42] Thus, large cities are conceived as being uniquely placed in the global economy to develop innovative climate solutions that straddle the public-private divide and serve as social laboratories for research and development. On a related note, C40 endorses the IPCC's findings that larger cities consume two-thirds of the world's energy and are responsible for more than three-quarters of global GHG emissions.[43] With this endorsement, C40 portrays its members as being responsible global stakeholders that aim to do something to reduce their contribution to climate change and that C40 itself plays a critical role by facilitating what the current chairman of C40, Eduardo Paes, calls 'city diplomacy'.[44] 'By engaging in city diplomacy, mayors and city officials

[37] Acuto, pg. 854.

[38] Greater London Authority, 'Mayor Brings Together Major Cities to Take Lead on Climate Change', online: www.london.gov.uk/media/mayor-press-releases/2005/10/mayor-brings-together-major-cities-to-take-lead-on-climate-change (accessed on 27 November 2016).

[39] C20 Climate Change Summit Communique (5 October 2005). [40] Ibid. [41] Ibid. [42] Ibid.

[43] C40, 'Ending Climate Change Begins in the City', online: www.c40.org/ending-climate-change-begins-in-the-city; IPCC, 'ADP Technical Expert Meeting: Urban Environment, Statement by Renate Christ, Secretary of the International Panel on Climate Change' (Bonn, 10 June 2014), online: www.ipcc.ch/pdf/unfccc/sbsta40/140610_urban_environment_Christ.pdf (both resources accessed on 29 November 2016).

[44] On C40's endorsement of the latest IPCC report, 'Climate Change 2014: Impacts, Adaptation and Vulnerability', which showed that many climate change risks are concentrated in urban areas, see C40, 'Mayor Paes, "IPCC Report Highlights Need for 'City Diplomacy' to Spur Climate Action"',

exchange information and experience. They facilitate the spread of new technologies and access to innovative public policies. Creative ideas and projects in one city can be replicated in another, and that exchange of knowledge is taking place, *far from lengthy and politically charged treaties*' (emphasis added).[45]

The communiqué sets out six specific actions:

Action #1: Commit to work together to set ambitious collective and individual, targets for reducing greenhouse gas emissions;
Action #2: Commit to ensure that we have highly effective agencies or programs dedicated to accelerating investments in municipal and community greenhouse gas emissions reductions and adaptation;
Action #3: Commit to develop, exchange, and implement best practices and strategies on emissions reductions and climate adaptation;
Action #4: Commit to develop and share communications strategies that sensitize citizens and stakeholders to climate change issues;
Action #5: Commit to create sustainable municipal procurement alliances and procurement policies that accelerate the uptake of climate friendly technologies and measurably influence the marketplace, including products containing greenhouse gases such as certain CFCs not covered by the UNFCCC;
Action #6: Meet again within 18 months in New York City to measure our progress and report back to the UN.
Source: C20 Climate Change Summit Communique (5 October 2005).

The 2007 New York City Summit communiqué, the 2009 Seoul Summit declaration, and the 2011 communiqué reiterate these aims – that is, to be a catalyst for climate action in cities around the world, to demonstrate that cities are taking up responsibility for climate change and are uniquely placed to craft innovation solutions, and to urge national governments to empower cities to undertake climate actions.[46]

5.3.2 *The Relationship with the Clinton Climate Initiative*

After C20 realized that its ambitions exceeded its institutional capacity, the network took steps to find a partner that it could work with to augment its resources. On 1 August 2006, C20 signed a MOU with the Clinton Foundation. Under the

online: http://c40.org/blog_posts/mayor-paes-ipcc-report-highlights-need-for-city-diplomacy (accessed on 30 November 2016).

[45] Eduardo Paes, 'C40 City Diplomacy Addresses Climate Change', *Huffington Post* (10 December 2013), online: www.huffingtonpost.com/eduardo-paes/c40-city-diplomacy-addresses-climate-change_b_4419898.html (accessed on 30 November 2016).

[46] The 2011 communiqué also served as a submission to the Rio+20 UN Conference on Sustainable Development as C40 wanted to highlight the link between cities' climate actions and wider sustainability goals (e.g. global environmental protection and poverty eradication); C40, 'C40 Cities Climate Leadership Group (C40) Communiqué and Resolution for the Rio+20 UN Conference on Sustainable Development' (3 June 2011), online: http://c40-production.herokuapp.com/blog_post - cities-climate-leadership-group-c40-communiqué-and-resolution-for-the-rio20-un-conference-on-sustainable-development (accessed on 30 November 2016).

MOU, the foundation's Clinton Climate Initiative (CCI) would serve as C20's delivery partner and 'utilize the global influence of President Clinton and the skills that it has developed in global mobilization to confront crises such as AIDS to help initiate programs that directly result in substantial reductions in [GHGs]'.[47] The MOU states that CCI's efforts would include, first, organizing a consortium that aggregates the purchasing power of cities to buy energy-saving products and technologies at lower prices.[48] This idea is similar to the Clinton Foundation's AIDS Initiative total quality management approach that has substantially lowered AIDS drug prices for members of its purchasing consortium.[49] Secondly, CCI would mobilize experts from the private sector to provide technical assistance to cities.[50] Thirdly, CCI would create and deploy common measurement tools and Internet-based communication systems to assist cities in creating emissions baselines, measuring the effectiveness of mitigation programmes, and sharing best practices.[51] Pursuant to the MOU, CCI also committed to raising funds to support the C20 agenda, which made cities like Melbourne perceive participation in C20 to be more beneficial and meaningful compared to involvement in other transnational city networks that lack the financial resources to undertake ambitious action.[52]

The Energy Efficiency Building Retrofit Program (EEBRP) exemplifies the mode of cooperation between C20 and CCI. As mentioned earlier, under the terms of the MOU, CCI committed to organizing a global procurement process that would enable C20 cities to purchase energy-saving technologies at preferential prices. CCI designed the EEBRP to operate as a 'matchmaking consortium'.[53] On the demand side, EEBRP gives

47 Memorandum of Understanding between the Large Cities Climate Leadership Group and the William J. Clinton Foundation, pg. 2; also see 'Press Release: President Clinton Launches Clinton Climate Initiative', 1 August 2006, online: www.clintonfoundation.org/main/news-and-media/press-releases-and-statements/press-release-president-clinton-launches-clinton-climate-initiative.html (accessed on 27 November 2016).

48 MOU, ibid.

49 Jeremy Youde, *Global Health Governance* (Polity Press 2012), pg. 85; Marilyn Chase, 'Clinton Foundation, Unitaid Strike Deals on Price Cuts for AIDS Drugs', 29 April 2008, online: www.wsj.com/articles/SB120943380712251563 (accessed on 27 December 2016).

50 MOU, pg. 1.

51 Ibid. The software tool, developed together with Microsoft and ICLEI, enables cities to implement a common measurement system for GHGs, access data from around the world, and conduct webinars with other cities. Known as HEAT, which stands for Harmonized Emissions Analysis Tool, this software was created by ICLEI and then further developed as an online tool by Microsoft and CCI; see ICLEI, 'ICLEI HEAT Software Goes International' (23 May 2007), online: www.iclei.org/details/article/iclei-heat-software-goes-international.html; and Microsoft News Center, 'Clinton Foundation, Microsoft to Develop Online Tools Enabling the World's Largest 40 Cities to Monitor Carbon Emissions' (17 May 2007), online: http://news.microsoft.com/2007/05/17/clinton-foundation-microsoft-to-develop-online-tools-enabling-the-worlds-largest-40-cities-to-monitor-carbon-emissions/ (both resources accessed on 29 December 2016).

52 City of Melbourne Council Meeting, *Agenda Item 5.3, C20: Large Cities Climate Leadership Group*, 4 July 2006, pg. 2, online: www.melbourne.vic.gov.au/AboutCouncil/Meetings/Lists/CouncilMeetingAgendaItems/Attachments/2006/EC_53_20060704.pdf (accessed on 27 December 2016).

53 Mikael Roman, 'Governing from the Middle: The C40 Cities Leadership Group' (2010) 10(1) *Corporate Governance: The International Journal of Business in Society* 73, pg. 76.

cities assistance to procure services for retrofitting their buildings for energy-saving and efficiency purposes.[54] On the supply side, companies that sell energy efficiency services and technologies are invited to supply the potentially large market of cities on the condition that they follow global best practices and are willing to provide their services at reduced prices.[55] The cities are under no obligation to buy from these companies, and final purchasing decisions are made independently of CCI.[56] As for the corporations, the terms of their participation in the EEBRP differ in each case.[57] Ultimately, the guiding principle is to create the market conditions that are conducive for enabling cities and corporations to work together to deliver transformative, large-scale retrofitting projects that have the potential to significantly reduce urban GHG emissions. The EEBRP has been a success. Within a few years, the consortium managed to initiate more than 250 projects to retrofit buildings for energy efficiency in twenty cities.[58]

The C20 soon expanded to include thirteen more cities and was renamed C40. However, as set out in the MOU with the Clinton Foundation, the group intended to continue as a 'small association of large and leading cities'.[59]

5.3.3 *Membership*

To become a member of C40, a city has to meet certain conditions. There are three types of membership categories:

Megacities (formerly Participating City):

Population: City population of 3 million or more, and/or metropolitan area population of 10 million or more, either currently or projected for 2025.

OR

GDP: One of the top twenty-five global cities ranked by current GDP output, at purchasing power parity, either currently or projected for 2025.

Innovator Cities (formerly Affiliate City):

Cities that do not qualify as Megacities but have shown clear leadership in environmental and climate change work.

An Innovator City must be internationally recognized for barrier-breaking climate work, a leader in the field of environmental sustainability, and a regionally recognized 'anchor city' for the relevant metropolitan area.

Observer Cities:

A short-term category for new cities applying to join the C40 for the first time; all cities applying for Megacity or Innovator membership will initially be admitted as Observers until they meet C40's first year participation requirements.

A longer-term category for cities that meet Megacity or Innovator City guidelines and participation requirements but, for local regulatory or procedural reasons, are unable to approve participation as a Megacity or Innovator City expeditiously.

Source: C40 membership categories[60]

[54] Ibid. [55] Ibid. [56] Ibid. [57] Ibid. [58] Roman, pg. 81. [59] MOU, pg. 1.

[60] Mike Marinello, *C40 Announces New Guidelines for Membership Categories* (C40 2012), online: c40-production-images.s3.amazonaws.com/press_releases/images/25_C40_20Guidelines_20FINAL_2011.14.12.original.pdf?1388095701 (accessed on 27 December 2016).

Only megacities are eligible for C40 governance and leadership positions – for example, serving as the C40 chair. C40 Megacities include Bangkok, Cairo, Chicago, Paris, London, and New York.[61] Innovator Cities include Oslo, Rotterdam, San Francisco, and Santiago, while Observer Cities include Beijing, Nairobi, Shanghai, and Singapore.[62]

To obtain as well as maintain its C40 membership, a city does not have to fulfil any formal performance-based obligations. However, as noted above, a city that is seeking C40 membership has to begin as an Observer City and is subject to what is effectively a one-year probation period. During this period and thereafter, a city is expected to demonstrate 'serious commitment' to addressing climate change and how it benefits from participating in C40.[63] The decision on whether a city obtains membership lies with the network's board of directors, which examines the city's track record and potential for ambitious climate change action as well as its ability to contribute to C40's work.[64]

C40 membership criteria are not made known to the public, but it is widely known that C40 has demanding standards because it wants to be selective and, to a certain extent, be an exclusive club.[65] C40 has indicated that in the long run, it plans to move towards clearly articulated and formal performance-based membership standards.[66] The Compact of Mayors is an agreement amongst city networks – including C40, ICLEI, and UCLG – which was launched at the 2014 UN Climate Summit.[67] As a signatory, C40 actively encourages its member cities to commit to the Compact of Mayors. Therefore, it is likely that in the near future, C40 cities will meet or exceed the compliance requirements of the Compact of Mayors. To commit to the Compact of Mayors, a city will first have to register on either of the Compact's reporting platforms: carbon*n* Climate Registry or CDP.[68] It should be noted that CDP is already the official reporting platform for C40 cities. Secondly, within

[61] C40, 'C40 Cities Make a Difference', online: www.c40.org/cities (accessed on 27 December 2016).

[62] Ibid. Singapore joined C40 as an observer city in March 2012. Singapore was invited to be part of the network as recognition of its good track record on balancing economic growth and sustainable development. As Singapore is a city state (and not a city within a larger, multi-level political system), it participates in C40 as an observer; Bouteligier, *Cities, Networks and Global Environmental Governance: Spaces of Innovation, Places of Leadership*, pg. 35.

[63] Interview No. 5. [64] Ibid. [65] Interviews Nos. 2, 7, 8, and 10.

[66] C40 and ARUP, *Climate Action in Megacities (C40 Cities Baseline and Opportunities, Volume 2.0, February 2014)* (2014), pg. 7.

[67] Launched at the Climate Summit in New York City on 23 September 2014, the Compact of Mayors is a high-profile initiative by the three major transnational city networks for climate change that exist today – C40, ICLEI, and UCLG – in partnership with UN-Habitat and Michael Bloomberg (in his capacity as the UN Secretary-General's Special Envoy for Cities and Climate Change); Compact of Mayors, online: www.compactofmayors.org (accessed on 1 July 2016).

[68] The carbon*n* Climate Registry is a global reporting platform launched in November 2010 at the World Mayors Summit on Climate in Mexico City to allow cities to report their goals and actions to the public; online: www.carbonn.org; CDP collects and discloses environmental information of major companies and cities around the world; online: www.cdp.net. The information in the rest of this paragraph is drawn from the 'Full Guide to Compliance for the Compact of Mayors' (July 2015), online: www.compactofmayors.org/history/ (accessed on 1 July 2016).

one year of its registration, a city must create a GHG inventory with a breakdown of emissions for the buildings and transportation sectors. This GHG inventory must be compiled using the GHG Protocol for Cities (see discussion in Section 5.3.5, 'Partnerships'). The city is also required to identify its climate risks. Both its inventory and risk profile must be reported via CDP or the carbon*n* Climate Registry. Within two years of its registration, the city must update its GHG inventory to include emissions from its waste sector. It also has to set a GHG reduction target and conduct a climate vulnerability assessment. Again, all the information must be made publicly available on one of the reporting platforms. Within three years of its registration, the city is required to deliver an action plan that shows how it will meet its GHG reduction target and its adaptation challenges. Upon completing all these requirements, a city will be certified 'compliant' and receive a certification logo that may be publicly displayed online and in printed materials. A new 'compliant' badge will be issued to the city each year that it maintains compliance through annual reporting.

C40 has the formal power to withdraw membership from cities that are deemed to be underperforming. However, asking a city to leave C40 is perceived to be a serious and drastic measure and has not yet been done.[69] Instead, the board of directors usually engages in informal discussions with the underperforming city, which is subsequently expected to take action towards improvement.[70] According to the *Climate Action in Megacities 2.0* survey, the global average is 137 actions per C40 member city. Based on responses received from fifty-three C40 cities, North American cities are the most active C40 members, as each city has undertaken, on average, 175 actions. East Asian cities have implemented the lowest number of climate initiatives (77 per city). Further, not all C40 cities have developed a climate change action plan with discernible or comparable quantitative targets. Shenzhen, for example, has a target of reducing its carbon dioxide emissions by 21 per cent per unit of GDP between 2010 and 2015, while Milan aims to reduce its carbon dioxide emissions by 20 per cent by 2020 (compared with its emissions in 2005). Sydney aims to reduce by 70 per cent by 2030 (compared with its emissions in 2006). Bangkok and Cairo, on the other hand, do not have emissions reduction targets.[71] The use of different years as a baseline and different ways of expressing the targets makes it difficult to compare the performance of cities and, as discussed throughout this book, has created the impetus to develop uniform reporting standards. As for cities which have yet to declare an emissions reduction target, they are considered the laggard cities that require much more capacity-building and resources to help them improve their climate performance.

[69] Interview No. 5. [70] Ibid.

[71] C40, online: http://c40-production.herokuapp.com/cities (accessed on 29 December 2016).

5.3.4 *Networks, Summits, and Workshops*

In establishing C40, London wanted to create a platform for cities to gain access to information about climate-related policies. Not having that knowledge itself when it wanted to go beyond national climate policies, London became keenly aware of how access to information is a crucial first step and how it could have saved much time if a network like C40 had existed when it needed such information.[72] Thus, from the outset, workshops, summits, and conferences take place on a regular basis to facilitate the exchange of ideas and best practices as well as build personal interactions (in addition to virtual ones).

The basic architecture of C40 is made up of issue-specific networks. C40 currently has sixteen networks in areas such as Adaptation and Water, Energy, Finance and Economic Development, Measurement and Planning, Solid Waste Management, Transportation, and Urban Planning and Development.[73] When a city becomes a member of C40, C40 staff will work with the city's representative to conduct a detailed analysis to determine the city's immediate priorities and preferences. Based on this analysis, the city will choose to join specific networks that cater to its interests – for example, in developing solid waste management solutions or improving the city's sustainable transportation options. A city will typically join four or five networks; it is discouraged from joining too many networks, as the assumption is that a city that is involved in too many networks will have its resources stretched too thinly to gain or contribute optimally.[74]

In the early stages, C40 selected a set of issues that served as the focal point of these networks. These issue areas are those in which city governments are most likely to have the legal powers to act such as waste management and energy efficiency in buildings.[75] Thus, there will not be a network on electricity generation because cities usually do not have the power to decide on the sources of their electricity. Networks have also emerged because of demand from member cities or at their initiative. There is a director who provides oversight of all the networks, in addition to a team of managers.[76] Each network has a manager that is responsible for its day-to-day operations, organizes meetings and workshops for the cities that are part of that network, promotes new ideas, and gathers resources to meet the needs of the member cities. In some networks, the manager also serves as the liaison person with external partners. For example, within the Adaptation and Water category, the C40 Cool Cities Network was launched in 2012 in partnership with Global Cool Cities Alliance, a non-profit organization that promotes research

[72] Bouteligier, pg. 92.
[73] C40, 'Networks: Connecting Cities on Topics of Common Interest', online: http://c40-production.herokuapp.com/networks (accessed on 30 December 2016).
[74] Interview No. 5. [75] Ibid.
[76] C40, 'Our Team', online: www.c40.org/our_team (accessed on 30 December 2016).

and policy awareness of solutions to reduce the urban heat island effect.[77] Members of the C40 Cool Cities Network include Tokyo, New York City, Athens, and Toronto.[78]

Once every two years, all C40 cities come together for the summit. A flagship event for the network, the C40 summit has been described as being 'like a low-key version of a heads of state summit'.[79] At the summits, mayors present their 'ground-breaking projects', forge strategic partnerships, and announce new initiatives to the public.[80] For example, in 2014, Johannesburg hosted the C40 summit during which the *Climate Action in Megacities 2.0 Report* was formally launched.[81] A quantitative survey of what C40 cities have done to reduce GHG emissions and improve climate resilience, the report provides a basis for measurement and evaluation. It also serves as a public relations tool to promote C40's message that 'cities have the power, the expertise, the political will and the resourcefulness to continue to take meaningful action'.[82] During the C40 summit in Johannesburg, Rotterdam mayor Ahmed Aboutaleb showcased his city's adaptation projects, such as the floating pavilion and the Stadshavens redevelopment project.[83] He also presented the book *Connecting Delta Cities: Resilient Cities and Climate Adaptation Strategies*, which showcases some of the activities and projects initiated under the auspices of Connecting Delta Cities, a small network of cities within C40's Water and Adaptation Initiative.[84] As mentioned in Chapter 4, Rotterdam is the founder of this network and operates a secretariat to support the network's activities. Finally, Mayor Aboutaleb also scheduled a meeting with the mayor of Durban to discuss potential opportunities in reducing GHG emissions from the maritime transport and logistics sectors.[85]

[77] C40, 'Cool Cities', online: www.c40.org/networks/cool_cities. On the urban heat island effect, see US EPA, 'Reducing Urban Heat Islands: Compendium of Strategies', online: www.epa.gov/sites/production/files/2014-06/documents/basicscompendium.pdf (both sources accessed on 30 December 2016).

[78] C40, ibid.

[79] Anita Powell, 'Mayors' C40 Summit Gives Megacity Leaders a Chance to Grab the Global Spotlight', *The Guardian*, 7 February 2014.

[80] C40 Summits: London Summit (2005), New York Summit (2007), Seoul Summit (2009), São Paolo Summit (2011), Johannesburg Summit (2014), Mexico City Summit (2016).

[81] C40 Blog, 'August Rewind: Plenary Videos from the C40 Mayors Summit', online: www.c40.org/blog_posts/august-rewind-plenary-videos-from-the-c40-mayors-summit (accessed on 1 June 2016).

[82] C40 Blog, 'Climate Action in Megacities Version 2.0', online: www.c40.org/blog_posts/CAM2 (accessed on 1 June 2016). This report shows that in the two years since C40 last surveyed its members in 2011, the total number of 'actions' taken by C40 cities has nearly doubled to 8,068 across a range of sectors including transportation, waste management, outdoor lighting, and water; C40 and ARUP, *Climate Action in Megacities (C40 Cities Baseline and Opportunities, Volume 2.0, February 2014)*, pg. 6.

[83] Rotterdam Climate Initiative, 'Rotterdam Mayor Aboutaleb to Johannesburg for Delta City C40 Mayors Summit', 31 January 2014, online:www.rotterdamclimateinitiative.nl/uk/news/rotterdam-mayor-aboutaleb-to-johannesburg-for-delta-city-c40-mayors-summit?news_id=2053 (accessed on 1 June 2016).

[84] Ibid. [85] Ibid.

The C40 workshops are more focused on specific themes of a more technical nature, such as reducing traffic congestion in cities and municipal waste treatment.[86] In November 2010, Hong Kong hosted a major international conference on climate change, Climate Dialogue: Low Carbon Cities for High Quality Living.[87] A C40 workshop was organized as part of the international conference, and its focus was on the enhancement of energy efficiency in buildings and green transportation (particularly the development of electric vehicles).[88] More than a hundred participants attended the closed-door workshop, including property developers, electricity generation companies, car manufacturers, research institutes, and officials from cities around the world.[89]

5.3.5 *Partnerships*

5.3.5.1 Partnering with the Private Sector

From the outset, C40 sought partnerships with the private sector and non-profit organizations like the Clinton Climate Initiative, which are 'committed to a business-oriented approach to climate change', to implement market-based solutions to climate change.[90] In this regard, C40 espouses a neo-liberal environmental approach.[91] These partnerships are crucial to C40's quest to be a catalyst for urban action on climate change. Many city authorities have limited resources, and the implementation of climate actions requires significant time, money, and human resources. The partnerships that C40 has forged help overcome the constraints that member cities face. For example, the MOUs signed with energy service companies (e.g. Siemens) and environmental consultancies (e.g. Arup)

[86] For example, Workshop on Transport and Congestion (London, December 2007), Deltas in Times of Climate Change Conference (Rotterdam, September 2010), Workshop on Low Carbon Cities for High Quality Living (Hong Kong, November 2010), Bus Rapid Transit Workshop (Jakarta, November 2013), and Solid Wastes Networks Workshop (Milan, October 2014).

[87] During this month, Hong Kong was abuzz with climate change conferences and events. Alongside this major international conference were numerous side events, including lectures by world-renowned climate scientists, receptions hosted by the US, British, and Swedish consulates, and a climate law conference at the University of Hong Kong. For details, see Climate Dialogue, 'Press Kit 06: Side Events', online: http://civic-exchange.org/materials/event/files/20101103-1106%2520Climate%2520Dialogue/20101103-1106_ClimateDialogue_SideEvents.pdf (accessed on 1 June 2016).

[88] Climate Dialogue, 'Press Kit 04: Conference Programme', online: http://civic-exchange.org/materials/event/files/20101103-1106%2520Climate%2520Dialogue/20101103-1106_ClimateDialogue_ProgrammeDetails.pdf (accessed on 1 June 2016).

[89] Ibid.

[90] Bruce Lindsey, CEO of the Clinton Foundation, quoted in Microsoft News Center, 'Clinton Foundation, Microsoft to Develop Online Tools Enabling the World's Largest 40 Cities to Monitor Carbon Emissions' (17 May 2007).

[91] Bouteligier, *Cities, Networks and Global Environmental Governance: Spaces of Innovation, Places of Leadership*, pg. 99.

allow C40 cities to purchase products and services at preferential prices.[92] Arup conducts UrbanLife Workshops for C40 cities that, to date, have focused on energy, waste, and water strategies. These workshops allow C40 cities to tap on Arup's technical expertise and policy consulting in developing climate action plans and programmes.[93]

As part of their collaboration, C40 and Siemens launched the City Climate Leadership Awards in 2013 to confer recognition on 'global cities demonstrating excellence in urban sustainability'.[94] In 2014, a seven-member panel consisting of architects, former mayors, representatives from the World Bank, C40, and Siemens nominated thirty finalists and selected ten winners.[95] An eleventh prize, the 'Citizen's Choice', was selected by public (online) vote. The awards process also serves as a platform for identifying and cataloguing innovative case studies that other cities can learn from.

5.3.5.2 Cooperation with Other Transnational Municipal Networks

C40 also cooperates with other transnational municipal networks to achieve its goals. The development of the Greenhouse Gas Protocol for Cities (GHG City Protocol) is a good illustrative example. Briefly, GHG accounting provides a detailed and replicable report of the GHG emissions generated by a specific actor. Just as financial accounts can be kept at the level of the project, firm, or country, GHG accounting can take place at different levels. GHGs are generally measured and reported at the national, firm, facility, or project levels.[96] In the late 2000s, C40 and ICLEI recognized that cities did not have an internationally accepted methodology for calculating and reporting GHG emissions at the city level. There were many standards available to cities, none of which were considered complete in their coverage. They differed in terms of what emission sources and GHGs are included in the inventory, how emissions sources are defined, and how

[92] Clinton Climate Initiative, 'President Clinton Announces Landmark Program to Reduce Energy Use in Buildings Worldwide' (New York, 16 May 2007), online: www.clintonfoundation.org/main/news-and-media/press-releases-and-statements/press-release-president-clinton-announces-landmark-program-to-reduce-energy-use.html (accessed on 30 December 2016).

[93] Arup, 'Tackling Climate Change with the C40', online: www.arup.com/Homepage_Archive/Homepage_C40.aspx (accessed on 30 December 2014).

[94] Siemens and C40 (joint press release), 'C40 & Siemens Kick Off 2014 City Climate Leadership Awards' (Munich/New York, 21 January 2014), online: www.siemens.com/press/en/pressrelease/?press=/en/pressrelease/2014/infrastructure-cities/ic201401006.htm&content[]=IC&content[]=CC&content[]=Corp (accessed on 30 December 2016).

[95] City Climate Leadership Awards 2014, online: http://www.c40.org/awards/2014-awards/profiles (accessed on 30 December 2016).

[96] For an example of project-level accounting, see Global Environmental Facility GHG accounting methodologies, online: www.thegef.org/gef/ghg-accounting (accessed on 1 June 2016). For discussion about the Greenhouse Gas Protocol which calculates and reports GHG emissions at the firm level, see Jessica Green, 'Private Standards in the Climate Regime: The Greenhouse Gas Protocol' (2010) 12 *Business and Politics Article* 3.

transboundary emissions are treated.[97] These inconsistencies made it difficult to compare the performance of cities, raised questions about the quality of the data that was available, and hindered the capacity of third parties to act as watchdogs.[98] Thus, C40 and ICLEI decided to work together to develop a globally accepted and harmonized standard such that 'by using the [protocol], cities will also strengthen vertical integration of data reporting to other levels of government, and should gain improved access to local and international climate financing'.[99]

In June 2011, C40 and ICLEI signed an MOU to begin developing the GHG City Protocol. A year later, the partnership was expanded to include World Resources Institute, the World Bank, UNEP, and UN-Habitat.[100] A draft of the GHG City Protocol was released in March 2012 for public comment. After the public comment period ended, the draft protocol was updated and then tested in thirty-five cities worldwide. The cities that pilot tested the GHG Protocol in 2013 included Kyoto, Tokyo, Buenos Aires, Lima, London, and Stockholm. During the same period, six in-person stakeholder consultation workshops were held in Beijing, São Paulo, London, Dar es Salaam, New Delhi, and Jakarta. Over 150 city officials, researchers, and practitioners provided feedback on the pilot GHG City Protocol. After revisions, the draft protocol went through a second round of public comment in July to August 2014. The final version of the GHG City Protocol was published in December 2014.[101] The GHG City Protocol will be further discussed in Chapter 6 as a type of voluntary standard.

5.3.5.3 C40–World Bank Partnership

Amongst C40's partnerships, its relationship with the World Bank is arguably its most significant one, as it gives C40 member cities access to the multilateral financial institution's resources, including financing for low-carbon infrastructure

[97] Greenhouse Gas Protocol, Global Protocol for Community-Scale Greenhouse Gas Emission Inventories, pg. 19, online: http://carbonn.org/fileadmin/user_upload/cCCR/GPC/GHGP_GPC.pdf (accessed on 1 July 2016).

[98] For discussion about the problems of incomplete and incompatible data collection methods, see Angel Hsu et al., 'Track Climate Pledges of Cities and Companies' (2016) 532 *Nature* 303.

[99] Greenhouse Gas Protocol, *Launch of First Global Standard to Measure Greenhouse Gas Emissions from Cities* (2014), online: www.ghgprotocol.org/Release/GPC_launch (accessed on 30 December 2016).

[100] Greenhouse Gas Protocol, 'About the GHG Protocol', online: www.ghgprotocol.org/about-ghgp (accessed on 30 December 2014). World Resources Institute is an environmental NGO that had worked with the World Business Council for Sustainable Development (a coalition of 200 multinational corporations committed to sustainable development) to develop an emissions accounting tool for companies known as the GHG Protocol Corporate Standard. This corporate standard has gained global acceptance, partly because of its adoption by the International Organization for Standardization (ISO) in 2006 as the basis for the ISO 14064-1 standard (*Specification with Guidance at the Organization Level for Quantification and Reporting of Greenhouse Gas Emissions and Removals*).

[101] This account of the process behind the final publication of the GHG Protocol standard is documented on the GHG Protocol website; online: http://ghgprotocol.org/GPC_development_process (accessed on 30 December 2016).

projects and its capacity-building programmes. The relationship between the World Bank and C40 can be conceptualized as one of orchestration, a term coined by Kenneth Abbott and his colleagues to refer to a situation whereby an international organization (the World Bank in the present case) supports and endorses an intermediary actor (C40 in the present case) to address target actors (global cities) in pursuit of the international organization's governance goals.[102] Orchestration is intended to be mutually beneficial. Intermediaries voluntarily participate in orchestration because they value the ideational and material support offered by the international organization. At the same time, international organizations engage in orchestration because the use of intermediaries helps overcome their resource constraints, which are significant relative to the demands of their governance tasks even in the case of a well-funded organization like the World Bank.

C40 and the World Bank began cooperation in 2009 within a wider partnership with Ecos (a Swiss environmental consultancy), the City of Basel, and the Swiss State Secretariat for Economic Affairs to develop the Carbon Finance Capacity Building Program.[103] Premised on the belief that cities in developing countries can benefit from the Kyoto Protocol's Clean Development Mechanism (CDM), the Carbon Finance Capacity Building Program aimed to increase the ability of cities to develop and implement CDM projects and therefore benefit from climate finance. The programme adopted a 'learning by doing' approach in four pilot cities. Dar es Salaam, Jakarta, Quezon City, and São Paolo were selected as pilot cities, and the CDM projects there served as case studies for other cities in developing countries.[104] Amongst the four cities, all are C40 members except Quezon City. It should therefore be noted that C40's collaboration with external partners generates benefits for cities outside the network too.

In June 2011, C40 deepened its collaboration with the World Bank. Both institutions announced the launch of a new partnership to address the two main structural issues that hindered cities from gaining access to carbon finance.[105] Pursuant to this partnership, C40 and the World Bank agreed to cooperate to develop a consistent approach for city climate action plans. At that time, there were standardized methodologies for climate action plans at the national level, but none existed for the city level. The lack of a standardized methodology for city-level climate action plans made it difficult for investors to assess the financial viability of urban climate

[102] See discussion on orchestration in Chapter 2 (Section 2.2.2).

[103] World Bank Institute, 'Carbon Finance Capacity Building Program', online: http://wbi.worldbank.org/wbi/about/topics/carbon-finance-capacity-building (accessed on 30 December 2016).

[104] World Bank Institute, 'Carbon Finance Capacity Building Program' (Washington, DC, 2011), online: http://wbi.worldbank.org/wbi/document/carbon-finance-capacity-building-program (accessed on 30 December 2016).

[105] C40 and the World Bank (joint press release), 'C40 and World Bank Form Groundbreaking Climate Change Action Partnership', online: www.c40.org/press_releases/press-release-c40-and-world-bank-form-groundbreaking-climate-change-action-partnership (accessed on 30 December 2016). The information contained in the rest of this paragraph is drawn from this source.

action proposals and therefore reluctant to fund them. C40 and the World Bank agreed to work together to develop a methodology so that cities would be better placed to attract private-sector investments for their low-carbon development programmes. C40 and the World Bank also committed to working together to develop a city-level GHG accounting protocol, and the World Bank was invited to join ICLEI and C40 to develop the GHG City Protocol. In addition, C40 agreed to identify and work with national governments and private-sector investors who are interested in providing project financing in C40 cities. The World Bank committed to identifying opportunities amongst sources of concessional finance, developing risk management instruments, and engaging the private sector via the International Finance Corporation.

The C40–World Bank partnership has grown in scale and ambition, culminating in two initiatives that were announced at the 2014 Climate Summit hosted by UN Secretary General Ban Ki Moon on 23 September 2014.[106] The Cities Climate Finance Leadership Alliance and the City Creditworthiness Partnership seek to help cities improve their creditworthiness, provide cities with technical assistance and transactional support, and catalyze capital flows, particularly the mobilization of private capital, to cities to increase investment in low-carbon and climate-resilient infrastructure.[107]

Through orchestration, C40 and World Bank have created a global–local relationship that facilitates cities in bypassing their national governments and directly engaging with other transnational actors when it comes to governing climate change. Through initiatives like the City Creditworthiness Partnership, C40 member cities have direct access to the World Bank's expertise in carbon finance and funding opportunities. For many cities, the access to such resources is vital, because climate financing is relatively new to many city governments, in developed and developing countries alike.[108] Many city government officials do not have sufficient

[106] The summit in New York was widely hailed as a success in galvanizing public opinion in favour of strong climate action ahead of the COPs in Lima and, more importantly, Paris, where the international community was expected to conclude an agreement that would include GHG emission reduction targets for developed and developing countries alike. For discussion, see Michael Jacobs, 'Five Ways Ban Ki-moon's Summit Has Changed International Climate Politics Forever' *The Guardian* (24 September 2014); 'Why Climate Change Is Back on the Agenda' *The Economist* (22 September 2014).

[107] Climate Initiatives Database, 'City Creditworthiness Partnership', online: http://climateinitiativesdatabase.org/index.php/City_Creditworthiness_Partnership. According to the World Economic Forum and World Bank, there is enormous unmet demand for investments in low-carbon and climate-resilient urban infrastructure in low- and middle-income countries. More than US$1 trillion per year is needed to finance the infrastructure gap, which is attributable to a number of reasons, including lack of capacity at the municipal level to formulate projects with adequate business plans and inadequate levels of municipal creditworthiness (which limit access to low-cost capital); Climate Summit 2014, 'The Cities Climate Finance Leadership Alliance Action Statement', pg. 3, online: www.un.org/climatechange/summit/wp-content/uploads/sites/2/2014/07/CITIES-Cities-Climate-Finance-Leadership-Alliance.pdf (both sources accessed on 13 December 2016).

[108] Interview No. 4.

knowledge about climate finance and do not know what funding opportunities for climate projects are available; in addition, there is a gap between the 'language used by investors' and that used by city governments.[109] City authorities therefore face obstacles in securing funding for large-scale climate mitigation and adaptation projects.[110] As partnerships such as that between C40 and World Bank bypass national governments and empower cities to be more financially independent, this has given rise to mixed responses on the part of national governments. Some national governments, South Africa being a good example, are keen to encourage their cities to issue green bonds and explore alternative sources of financing to pursue low-carbon development.[111] Other national governments are less enthusiastic because of political reasons and financial risks.[112] Given that these developments are very recent and unfolding rapidly, it is too early to assess how states react to being bypassed by orchestration efforts except to note that we can expect a range of responses, which are highly country-specific, and that the current trend of building capacity for urban climate finance will continue apace for some time because of the momentum created by the mobilization of vast resources from international financial institutions, development agencies, supranational authorities like the European Union, and the private sector.[113] At the UNFCCC level, Executive Secretary Christiana Figueres has shown appreciation for the benefits of orchestration efforts and embraces the notion that polycentric, multi-actor governance ultimately reinforces and strengthens the transnational climate change regime complex.[114] In Chapter 6, the complementarity between urban transnational climate change governance and the UNFCCC regime will be further explored from the rule-making perspective.

[109] Ibid. [110] Interview No. 2. [111] Ibid.

[112] When a mayor is a well-known individual with potential aspirations at the national level of politics, his/her city's attempt to finance high-profile projects with international sources may cause tensions with the central government. A recent example is the city of Dakar, which was ready in February 2015 to issue the first municipal bond in West Africa after many years of preparation, but the issuance was blocked by the Senegalese government; see Sam Barnard, *Climate Finance for Cities: How Can International Climate Funds Best Support Low-Carbon and Climate Resilient Urban Development?* (Overseas Development Institute, Working Paper 419, 2015), pg. 20.

[113] Since the launch of the Cities Climate Finance Leadership Alliance, there has been a proliferation of urban climate finance capacity-building programmes and related initiatives. A working group under the auspices of the Cities Climate Finance Leadership Alliance has been put together to carry out a mapping exercise of all known urban climate financing initiatives (Interview No. 2). One of the latest initiatives, launched in early 2016, is the Low Carbon City Lab (LoCaL)'s Matchmaker programme, which aims to connect investors with urban mitigation projects; online: http://local.climate-kic.org/projects/matchmaker/ (accessed on 20 June 2016).

[114] This view finds resonance with orchestration theory rather than regime complex theory, which tends to see such multiplicity as a problem that threatens governance effectiveness through incoherence, redundancy, and conflict. See the address by Christiana Figueres, Executive Secretary of the UNFCCC, at the C40 Cities Mayor Summit in Johannesburg, 5 February 2014, online: http://unfccc.int/files/press/statements/application/pdf/20140502_c40_check.pdf (accessed on 1 June 2016).

5.4 CONCLUSION

This chapter has examined how global cities, in seeking to scale up their climate governance efforts and learn from each other, have found it beneficial to form a network. C40 is not only a horizontal network that connects city officials around the world; it has also created significant linkages to private actors such as global technology companies and engineering firms, prominent foundations like the Clinton Foundation, and international organizations including UN-Habitat and the World Bank. These linkages serve to connect cities directly to global actors to engage in climate governance. Through the C40 network, cities are also able to develop standards and practices that are uniquely suited to meet their needs and priorities, such as the GHG City Protocol. When standards and norms are created by cities and implemented through transnational networks, it can be argued that a specific form of governance – lawmaking – has taken place. Chapter 6 will make the argument that cities, through networks like C40, are beginning to play a lawmaking role in transnational climate change governance.

6

Cities as Transnational Lawmakers

6.1 INTRODUCTION

A key theme that has emerged from the foregoing discussion is that global cities are sites of innovative climate governance. Importantly, global cities are not just aiming to reduce GHG emissions locally. In line with the emerging trend of cities playing an increasingly visible role in international affairs, global cities aim to scale up their climate actions to generate worldwide impact and play a role in governing climate change at the global level. This has led to the formation of numerous city networks, of which the most prominent transnational network of global cities focusing on climate action is C40. Through reiterative interaction and frequent cooperation within their network, global cities develop and internalize certain norms, defined as rules that set 'a standard of appropriate behaviour for actors with a given identity'[1] and are imbued with a quality of 'oughtness' that sets them apart from other kinds of rules.[2]

In the present case, the norms are as follows: first, that climate change is a global problem but *can* and *must* be addressed locally by cities; second, that large, global cities are not only a source of the problem because of their high levels of GHG emissions but also a source of solutions; third, that cities can best reduce their GHG emissions and embark on low-carbon growth by committing to a policy approach based on measurement, monitoring, and reporting of their GHG emissions. Based on this normative foundation, global cities have cooperated (through C40) with other actors such as GHG accounting consultancies, development banks, and civil society to develop practices and voluntary standards to enable global cities to reduce

[1] Martha Finnemore and Kathryn Sikkink, 'International Norm Dynamics and Political Change' (1992) 52 *International Organization* 887, pg. 891.

[2] What is 'appropriate' is determined by a community or a society, which raises the question of how many actors must share in the collective assessment of a rule before it can be called a 'norm'; M. Finnemore and K. Sikkink suggest that, based on empirical studies, we can expect a norm to reach tipping point when at least one-third of the total states in the system adopt the norm. Further, which states adopt the norm is an important factor. 'Critical states' are those whose non-adoption will compromise the achievement of the substantive norm; ibid., pg. 901. See discussion in Section 6.3 on norm diffusion.

their GHG emissions in the short term and make the transition towards low-carbon development in the longer term. I refer to these norms, practices, and voluntary standards developed by global cities and implemented through their transnational networks as *urban climate law*.

Urban climate law is notable in at least three respects. From the viewpoint of environmental effectiveness and fulfilling one of the key objectives of the transnational climate change regime complex (i.e. climate mitigation), urban climate law can make an important contribution towards reducing cities' GHG emissions, which constitute a sizeable share of global emissions (estimated to be 37–49 per cent).[3] From a theoretical perspective, understanding how urban climate law fits within the broader transnational climate change governance landscape and interacts with other normative institutions, particularly the UNFCCC regime, provides us with novel insights about the lawmaking role that cities have in the transnational climate change regime complex. The study of urban climate law in this chapter can make a contribution to the body of literature on 'soft law', specifically voluntary standards which tend to occupy a central position in the world of soft law solutions. Voluntary standards are characterized by the voluntary participation of actors in the construction and implementation of norms and practices. Participants are free to leave the voluntary scheme any time, and there is an absence of 'police power as a way to induce consent and compliance'.[4] In the past two decades, industry and governments have increasingly turned to voluntary schemes to address social and environmental externalities. There has therefore been a proliferation of voluntary schemes that create codes of conduct, standards, and indicators to address issues ranging from deplorable labour conditions in the global garment industry to the deleterious environmental effects of large-scale biofuels production.[5] This chapter argues that urban climate law constitutes a novel type of voluntary standard, and the discussion found here seeks to expand the analytical discourse of voluntary standards by considering the role of sub-national actors in creating and implementing voluntary standards.

To understand how urban climate law can lead to cities reducing their GHG emissions, this chapter identifies two key pathways of influence that are critical for shaping how cities view their interests and align them with the overarching

[3] Cambridge Institute for Sustainability Leadership, Cambridge Judge Business School, and ICLEI, *Climate Change: Implications for Cities (Key Findings from the Intergovernmental Panel on Climate Change Fifth Assessment Report)* (2014).

[4] John Kirton and Michael J. Trebilcock, 'Introduction: Hard Choices and Soft Law in Sustainable Global Governance' in John Kirton and Michael J. Trebilcock (eds), *Hard Choices, Soft Law: Voluntary Standards in Global Trade, Environment and Social Governance* (Routledge 2004), pg. 9.

[5] For discussion, see for example, Arthur P. J. Mol, 'Environmental Authorities and Biofuel Controversies' (2010) 19 *Environmental Politics* 61; Charan Devereaux and Henry Lee, *Biofuels and Certification: A Workshop at Harvard Kennedy School* (Discussion Paper 2009–07, Cambridge, MA: Belfer Center for Science and International Affairs 2009); Allison Loconto, 'Assembling Governance: The Role of Standards in the Tanzanian Tea Industry' (2015) 107 *Journal of Cleaner Production* 64; Ethical Trading Initiative Base Code, online: www.ethicaltrade.org (accessed on 1 July 2016).

objectives of reducing GHG emissions in the short term and developing low-carbon alternatives for the future. These pathways of influence involve the *promotion of reflexivity* in cities and *norm diffusion*. At the heart of the concept of reflexivity is that actors constantly reflect upon their social practices and have the capacity to make adjustments to those practices in light of new information.[6] In this line of thinking, voluntary standards are regulatory tools that uncover new information for cities, which can then reflect and act upon the information by adjusting their practices accordingly. Such information will identify, for example, the most cost-efficient climate mitigation opportunities available to a city.[7] The second pathway of influence is concerned with how the norms of urban climate law are disseminated and become widely adopted within a fairly short period of time. I refer to prominent theories of diffusion to offer an account of how urban climate law reaches a 'tipping point', after which it 'cascades' through the transnational city network and leads to climate action.[8]

As for the question of how urban climate law fits within the broader transnational climate change governance landscape and relates to the UNFCCC regime, this chapter advances the claim that, in many key respects, urban climate law has been deliberately designed to support and reinforce the UNFCCC regime. In fact, urban climate law interacts with the UNFCCC rules and institutions in strategic ways that are mutually reinforcing. I use the term 'coupling' to refer to the deliberate effort to align urban climate law in ways such that its norms and practices complement and strengthen the UNFCCC regime. The first example of coupling involves cities *reframing the issue* of climate change in ways that sidestepped the contentious issues that obstructed the UNFCCC negotiations and created gridlock. By reframing the problem and using an alternative framework of ideas as a basis for developing urban climate law, cities created a set of institutions that complemented the UNFCCC regime by providing less-controversial ways of promoting GHG reductions and reaching actors that, depending on their location, were not bound by the Kyoto Protocol. In the second example of coupling, urban climate law reinforces the UNFCCC regime by serving as a means for *diffusing the norms* underpinning the UNFCCC framework. I illustrate this point by reference to C40's adherence to the principle of common but differentiated responsibilities and respective capabilities (CBDRRC), which is a cornerstone of the international climate regime. My third

6 See discussion in Section 6.3.1.

7 See, for example, the Low Carbon City Development Program Assessment Protocol, designed to help cities plan, implement, monitor, and account for low-carbon investments and mitigation actions; World Bank, 'The Low Carbon City Development Program (LCCDP) Guidebook: A Systems Approach to Low Carbon Development in Cities (English)', online: http://documents.worldbank.org/curated/en/2014/01/24089839/low-carbon-city-development-program-lccdp-guidebook-systems-approach-low-carbon-development-cities (accessed on 1 July 2016).

8 For example, Finnemore and Sikkink; Zachary Elkins and Beth A. Simmons, 'On Waves, Clusters and Diffusion: A Conceptual Framework' (2005) 598 *Annals of the American Academy of Political and Social Science* 33.

example is the Compact of Mayors, which deliberately replicates features of the post-2020 international climate agreement, especially the pledge-and-review approach and monitoring, reporting, and verification (MRV). In this case, coupling also allows cities to serve as social laboratories for generating experience and knowledge of new norms and practices that are emerging from the Paris Agreement.

I argue that coupling not only benefits the UNFCCC regime but also supports the development of urban climate law by conferring legitimacy by association and implicit endorsement by the UNFCCC regime. The analysis on coupling also responds to the broader criticisms against soft law by demonstrating that the soft law generated by cities is an important complement to 'hard law' that can facilitate experimentation at multiple sites and levels of governance, generate knowledge (through exchange of good practices, for example), build trust, and transform norms.[9] Further, soft law and hard law not only support and complement one another, but the interaction between them can result in an overall expansion of governance and authority.

This chapter proceeds in five parts. Following this introduction, Section 6.2 provides a brief overview of the literature on soft law and voluntary standards. Section 6.3 advances the argument that, through their networks, cities develop and implement law that follows certain pathways of influence that can eventually lead to the reduction of GHG emissions. These pathways of influence involve the *promotion of reflexivity* in cities and *norm diffusion*. In Section 6.4, I argue that cities play an important role in strengthening the UNFCCC regime's normative influence. Urban climate law interacts with the UNFCCC rules and institutions in strategic ways that are mutually reinforcing. As a result, there is not only a lack of conflict between the two sets of rules; there is, in fact, complementarity that contributes to the coherence of the transnational climate change regime complex. Section 6.5 draws a number of conclusions on the significance of urban climate law and the role of cities as rule-makers.

6.2 SOFT LAW AND VOLUNTARY STANDARDS

6.2.1 *Soft Law versus Hard Law*

In the classical international law tradition, international legal rules are taken to be those that flow from the formal sources identified in Article 38 of the Statute of the International Court of Justice. As such, hard law would comprise the rules created by

[9] This line of argument finds commonality with those of constructivist scholars who focus less on the binding nature of law at the enactment stage and more on the effectiveness of law at the implementation stage, and who stress how soft law can 'facilitate constitutive processes such as persuasion, learning, argumentation and socialization', David M Trubek, Patrick Cottrell, and Mark Nance, 'Soft Law, Hard Law and EU Integration' in Gráinne de Búrca and Joanne Scott (eds), *Law and New Governance in the EU and the US* (Hart Publishing 2006), pg. 75.

international conventions, international custom, and 'general principles of law recognized by civilized nations'.[10] The term 'soft law' is used to refer to legally binding instruments 'which are only softly enforced (for example, with no courts to resort to) as well as instruments which are in the grey zone of normativity, be they softly binding in some respects only, or in the process of becoming law as part of the formation of customary international law'.[11] Thus, soft law would include a range of international instruments and communications ranging from informal understandings or conversations to memoranda of understanding, diplomatic letters, protocols, codes of conduct, and informal agreements.[12]

This broad overview of hard law and soft law glosses over the considerable disagreement in the literature on the definition of soft law and what falls into this category. The rules of recognition of international law are not bright line rules and are becoming more vague as 'international law is increasingly seen as a continuum between law and non-law, with formal law-ascertainment no longer capable of capturing legal phenomena in the international arena'.[13] Some international legal scholars use a simple 'binding/non-binding' binary to distinguish hard from soft law.[14] From a positive legal scholar's viewpoint, the concept of soft law is logically flawed because law is, by definition, of a binding nature, and thus there cannot be non-binding law.[15] Proper Weil, for example, laments the 'blurring of the normativity threshold' and argues that 'the threshold does exist: on one side of the line, there is born a legal obligation that can be relied on by a court or arbitrator, the flouting of which constitutes an internationally wrongful act giving rise to international responsibility; on the other side, there is nothing of the kind'.[16]

Constructivist scholars, in contrast, focus less on the binding nature of law. They are more interested in how rules actually operate in practice. For these scholars,

[10] Article 38(1), the Statute of the International Court of Justice. It should be noted that this list of not exhaustive. Article 38(2) states, 'This provision shall not prejudice the power of the Court to decide a case *ex aequo et bono*, if the parties agree thereto.'

[11] Joost Pauwelyn, 'Is It International Law or Not, and Does It Even Matter?' in Joost Pauwelyn, Ramses Wessel, and Jan Wouters (eds), *Informal International Lawmaking* (Oxford University Press 2012), pg. 129.

[12] Christine Chinkin, 'Normative Development in the International Legal System' in Dinah Shelton (ed), *Commitment and Compliance: The Role of Non-Binding Norms in the International Legal System* (Oxford University Press 2003), pgs. 25–31.

[13] Jean d'Aspremont, 'From a Pluralization of International Norm-Making Processes to a Pluralization of Our Concept of International Law' in Joost Pauwelyn, Ramses Wessel, and Jan Wouters (eds), *Informal International Lawmaking* (Oxford University Press 2012), pg. 195.

[14] See Jan Klabbers, 'The Redundancy of Soft Law' (1996) 65 *Nordic Journal of International Law* 167 for argumentation in favour of retaining the binary conception of law. Wolfgang H. Reinicke and Jan Martin Witte, 'Interdependence, Globalization, and Sovereignty: The Role of Non-Binding International Legal Accords' in Dinah Shelton (ed), *Commitment and Compliance: The Role of Non-Binding Norms in the International Legal System* (Oxford University Press 2003), pgs. 75–76.

[15] Jan Klabbers argues that law cannot be 'more or less binding', and therefore the concept of soft law is logically flawed; ibid., pg. 181.

[16] Prosper Weil, 'Towards Relative Normativity in International Law' (1983) 77 *American Journal of International Law* 413, pgs. 415, 417–418.

what is salient is whether and how those who adhere to norms, rules, and standards come to accept and regard them as authoritative.[17] Brunnée and Toope, for example, argue, 'We should stop looking for the structural distinctions that identify law, and examine instead the processes that constitute a normative continuum bridging from predictable patterns of practice to legally required behavior.'[18] Constructivist scholars also argue that formally binding rules are not necessarily more effective than rules of a non-binding nature. As Kal Raustiala points out, we should distinguish between effectiveness and compliance:

> Compliance as a concept draws no causal linkage between a legal rule and behavior, but simply identifies a conformity between the rule and behavior. To speak of effectiveness is to speak directly of causality: to claim that a rule is 'effective' is to claim that it led to certain behaviors or outcomes, which may or may not meet the legal standard of compliance.[19]

The definition of legalization in international relations proposed by Kenneth Abbott and Duncan Snidal offers another alternative for understanding the distinction between hard law and soft law without resorting to the narrow conceptualization offered by classical international law and the binding/non-binding dichotomy. Abbott and Snidal define legalization in international relations as varying across three dimensions – (i) precision of rules, (ii) obligation, and (iii) delegation to a third-party decision-maker.[20] Hard law 'refers to legally binding obligations that are precise (or can be made precise through adjudication or the issuance of detailed regulations) and that delegate authority for interpreting and implementing the law'.[21] International trade law is said to come closest to this ideal type of hard law.[22] In Abbott and Snidal's definition, 'the realm of "soft law" begins once legal arrangements are weakened along one or more of the dimensions of obligation, precision, and delegation. This softening can occur in varying degrees along each dimension and in different combinations across dimensions.'[23] They emphasize that soft law is a shorthand term to refer to this 'broad class of deviations from hard law' and that soft law comes in many varieties.[24]

Some scholars have questioned the characterization of law in terms of these three attributes (precision, obligation, and delegation) on the basis that law encompasses

17 Trubek, Cottrell, and Nance, pgs. 80–81.

18 Jutta Brunnee and Stephen J. Toope, 'International Law and Constructivism: Elements of an Interactional Theory of International Law' (2000) 39 *Columbia Journal of Transnational Law* 19, pg. 68.

19 Kal Raustiala, 'Compliance and Effectiveness in International Regulatory Cooperation' (2000) 32 *Case Western Reserve Journal of International Law* 387, pg. 398.

20 Kenneth W. Abbott and Duncan Snidal, 'Hard and Soft Law in International Governance' (2000) 54 *International Organization* 421, pg. 424.

21 Ibid., pg. 421.

22 Gregory Shaffer and Mark A. Pollack, 'Hard vs. Soft Law: Alternatives, Complements, and Antagonists in International Governance' (2010) 94 *Minnesota Law Review* 706, pg. 715.

23 Abbott and Snidal, pg. 422.

24 Ibid.

more than these largely technical and formal criteria.[25] However, for present purposes, Abbott and Snidal's definition is helpful in terms of delineating some 'indicators' of legalization which, in turn, define a continuum on which international treaty rules, voluntary rules, and codes of conduct can be situated. From this viewpoint, we can say that states, private actors, and civil society have increasingly used a wide range of normative instruments that have a relatively harder or softer legal nature in terms of precision, obligation, and delegation to pursue their objectives.

6.2.2 *Voluntary Standards*

Standards are norms selected as a model by which people, actions, or products can be judged and compared, and which provide a common language for the judges, the evaluated, and their audiences.[26] Standards address technical and compatibility issues by operating as tools of simplification and specification.[27] Voluntary standards can be found across many domains of contemporary economies, including tourism, construction, and mining.[28] In recent years, voluntary standards have gained prominence as a means of addressing a product or an industry's impact on the natural environment. In forestry, for example, voluntary standards have filled the governance gap caused by the inability of governments to come to agreement on how best to protect forests from unsustainable use.[29] The increasing popularity of voluntary standards has generated an extensive literature examining the effectiveness of voluntary standards,[30] their legitimacy,[31] and their potential to work in tandem with

[25] See, for example, Martha Finnemore and Stephen J. Toope, 'Alternatives to "Legalization": Richer Views of Law and Politics' (2001) 55 *International Organization* 743.

[26] Stefano Ponte and Emmanuelle Cheyns, 'Voluntary Standards, Expert Knowledge and the Governance of Sustainabilty Networks' (2013) 13 *Global Networks* 459, pg. 461.

[27] F. Ewald, 'Insurance and Risk' in G. Burchell, C. Gordon and P. Miller (eds), *The Foucault Effect: Studies in Governmentality* (Harvester 1990).

[28] Green Globe's International Standard for Sustainable Tourism (Version 1.7) is a certification scheme that requires tourism agencies and businesses to report on their sustainability performance throughout their supply chains; see Green Globe, online: http://greenglobe.com/green-globe-certification/. Leadership in Energy and Environmental Design (LEED) is a third-party certification scheme for 'green buildings'; LEED, online: www.usgbc.org/articles/about-leed. Fairtrade Gold and Precious Metals is a voluntary certification scheme that requires small-scale mining organizations and downstream operators to meet standards pertaining to responsible environmental management, labour conditions, and women's rights; online: www.fairgold.org (all Internet links accessed on 15 July 2016).

[29] For discussion, see for example, Benjamin Cashore et al., 'Forest Certification in Developing and Transitioning Countries: Part of a Sustainable Future?' (2006) 48 *Environment* 6.

[30] For example, in response to the question of whether fair trade standards improve the livelihoods of coffee farmers, research has shown that the fair trade price does improve the welfare of certified growers, but the higher price alone does not address the broader challenges that these marginalized farmers face; K. Utting-Chamorro, 'Does Fair Trade Make a Difference? The Case of Small Coffee Producers in Nicaragua' *Development in Practice* 15: 3/4 (2005), pp. 584–599.

[31] See, for example, Graeme Auld and Lars H. Gulbrandsen, 'Transparency in Nonstate Certification: Consequences for Accountability and Legitimacy' (2010) 10 *Global Environmental Politics* 97; Steven Bernstein and Benjamin Cashore, 'Can Non-State Global Governance Be Legitimate? An Analytical Framework' (2007) 1 *Regulation & Governance* 347.

'relatively hard' institutions to achieve ecologically and socially sustainable practices in various sectors, such as forestry, shipping, and food production.[32]

Before proceeding further to consider some of the voluntary standards literature, I like to note that a significant amount of the research on voluntary standards is premised on the public-private dichotomy.[33] According to this public-private distinction, only states (and international organizations, which states create and delegate authority to according to the traditional principal–agent model) have the authority to make *mandatory* rules. All other actors belong to the catch-all category known as 'non-state actors' or 'private actors'; these have no authority to make mandatory rules and therefore create *voluntary* rules.

The distinction is blurred when we expand our conception of rule-making in the global sphere to include cities as rule-makers. Cities are public in the sense that they exercise functions of public administration but are not states. They are also not private in the same way that business corporations and NGOs are. For example, cities do not share the profit maximization goal that firms do by dictates of law and culture. Further, cities are not non-state actors. Cities are sub-state actors. They are not just physically embedded within the territory of states, but are also subject to the authority of central governments. Cities do not have authority to make mandatory rules at the international level, but it does not follow that cities are therefore private/non-state actors. In brief, the public-private distinction does not capture the reality of contemporary transnational lawmaking in which the city plays a part because of a definitional conflation. The public-private definition conflates municipal authorities, provincial governments, businesses, environmental NGOs, labour unions, and private foundations into a single category. Yet, as I have suggested, cities have interests and shared ideas about their identity that are distinctly different from those of, for example, multinational corporations. I therefore extend existing definitions and theoretical conceptions of voluntary standards in the discussion that follows to clarify the point that voluntary standards can be created by non-state actors *as well as* sub-state actors such as cities.[34] This conceptual move aims to open

[32] See, for example, Guy Salmon, *Voluntary Sustainability Standards and Labels (VSSLs): The Case for Fostering Them* (Roundtable on Sustainable Development, OECD, 2002), online: www.oecd.org/sd-roundtable/papersandpublications/39363328.pdf (accessed on 15 July 2016).

[33] For example, Abbott and Snidal; Jessica Green, 'Private Standards in the Climate Regime: The Greenhouse Gas Protocol' (2010) 12 *Business and Politics Article* 3; Tim Buthe and Walter Mattli, *The New Global Rulers: The Privatization of Regulation in the World Economy* (Princeton University Press 2011); David Vogel, 'Trading Up and Governing Across: Transnational Governance and Environmental Protection' (1997) 4 *Journal of European Public Policy* 556. Within the discipline of international relations, by and large, 'the public' refers to the state, while 'the private' refers to the non-state; Philipp Pattberg and Johannes Stripple, 'Beyond the Public and Private Divide: Remapping Transnational Climate Governance in the 21st Century' (2008) 8 *International Environmental Agreements* 367, pg. 372.

[34] I recognize that scholars like Abbott and Green acknowledge the role of sub-national governments in voluntary rule-making, but they either subsume cities and other local governments within the category of 'non-state actors' or identify the transnational municipal network (e.g. C40) as the relevant

up the analytical discourse to consider the role of sub-state actors in creating voluntary standards and transnational law more broadly.

For example, in Abbott and Snidal's concept of 'regulatory standard setting' – which is said to occur when voluntary standards are adopted by firms, NGOs, and states on their own or in partnerships involving at least two out of the three actors – the category of NGOs contains a diverse group of actors.[35] It includes advocacy groups, labour unions, social movements, and 'other noncommercial groups'.[36] I suggest that, while cities can be accommodated within the 'other noncommercial groups' sub-grouping, it would be more apposite to specify the city as an actor with a unique contribution to make towards regulatory standard setting. Nonetheless, for present purposes, it suffices to note that Abbott and Snidal's theory of regulatory standard setting aptly captures the rule-making activities of cities through the transnational networks they create. More specifically, cities can be regarded as *transnational regulators*.

Green coined the term 'entrepreneurial private authority' to refer to voluntary standards. She specifies that non-state actors make these rules and set these standards 'without the explicit delegation of authority by states'.[37] A number of implications follow from Green's definition that are salient in constructing our understanding of urban climate law. First, entrepreneurial private authority is restricted to instances in which non-state and sub-state actors create rules, standards, and practices that govern the conduct of others. According to Green, this means that operational activities such as capacity building, information sharing, and the publication of action plans are not instances of entrepreneurial private authority, as these activities do not prescribe rules.[38] However, I adopt a different stance. While these operational activities do not constitute rule-making, they are important ways in which those who seek to govern persuade other actors to follow their rules. Furthermore, these operational activities fall within the broader ambit of governance and usually have normative content. I therefore consider the operational activities that global cities

actor. My aim in extending their definitions is to clarify that cities are not non-state actors and to emphasize that cities are the actors making and implementing these rules.

35 Abbott and Snidal point out that their use of the term 'regulatory standard setting' is intended to highlight that these voluntary standards adopted by firms, states, and other actors seek to go beyond meeting demands for technical coordination and to address social and environmental externalities; Kenneth W. Abbott and Duncan Snidal, 'The Governance Triangle: Regulatory Standards Institutions and the Shadow of the State' in Walter Mattli and Ngaire Woods (eds), *The Politics of Global Regulation* (Princeton University Press 2009), pg. 45. These voluntary standards are therefore regulatory in nature, with regulation defined as 'the organization and control of economic … and social activities by means of making, implementing, monitoring, and enforcing of rules'; Walter Mattli and Ngaire Woods, 'In Whose Benefit? Explaining Regulatory Change in Global Politics' in Walter Mattli and Ngaire Woods (eds), *The Politics of Global Regulation* (Princeton University Press 2009).

36 Ibid., pg. 60.

37 Jessica Green, *Rethinking Private Authority: Agents and Entrepreneurs in Global Environmental Governance* (Princeton University Press 2013), pg. 78.

38 Ibid., pg. 30.

engage in to be important and relevant for creating an institutional environment that is conducive for the development and implementation of urban climate law. The second implication of Green's definition is that those who wish to govern have to persuade others to follow their rules in order to exercise authority. In Section 6.2 of this chapter, I will analyze how certain leading cities and their mayors engaged in persuasion to convince other actors to follow their lead. Thirdly, the claimant to authority relies on expertise to legitimize its claim and is subsequently able to induce behavioural change in some relevant actor in global affairs.[39]

Finally, to round up the present discussion on voluntary standards, it is noteworthy that voluntary standards are so pervasive that their existence has been acknowledged by the World Trade Organization (WTO).[40] The WTO's Technical Barriers to Trade Agreement (TBT Agreement) draws a distinction between regulations and standards. While compliance with the former is mandatory, compliance with the latter is voluntary.[41] Drawing upon this, and upon discussions in the WTO's Sanitary and Phytosanitary Measures committee on voluntary standards, Scott defines 'private standards' as 'written documents adopted by a non-governmental entity that lay down rules, guidelines and/or characteristics, for common or repeated use, for products or related processes and production methods, including transport'.[42] Again, urban climate law comes within Scott's definition of private standards, save that the term 'non-governmental entity', as I have consistently argued, ought to be disaggregated to refer to both non-governmental and sub-state entities.

6.3 THE WORKINGS OF URBAN CLIMATE LAW

'Urban climate law' refers to the norms, practices, and voluntary standards constructed by cities which, through certain pathways, are transmitted transnationally and adopted by cities across the world. Before proceeding further to examine the pathways by which urban climate law steers cities towards climate mitigation and long-term decarbonization strategies, this section will provide a brief overview of the normative framework of urban climate law and some of the voluntary standards that have been developed.

As mentioned earlier, one can discern three key normative ideas that global cities and their networks have sought to promote. These normative ideas underpin the attempts by global cities to claim a role in transnational climate change governance. Briefly, the three norms are these: first, that climate change is a global problem but *can* and *must* be addressed locally by cities; second, that large, global cities are not only a source of the problem because of their high levels of GHG emissions but also

39 Ibid., pg. 36.

40 Joanne Scott, 'The Promise and Limits of Private Standards to Reduce Greenhouse Gas Emissions from Shipping' (on file with author), pg. 8.

41 Ibid. 42 Ibid.

a source of solutions because of their concentration of human capital and economic resources (cities are also perceived to be more nimble political actors than national governments); third, that the optimal approach for cities to reduce their GHG emissions and embark on low-carbon growth is one based on data transparency (this involves cities regularly monitoring, measuring, and reporting their GHG inventories and the impact of the climate actions they have taken).

Based on this set of norms, cities have worked with other actors to develop voluntary standards and related practices. The Compact of Mayors is an example of a voluntary standard. As I briefly described in Chapter 5, C40 and other city networks came together to create the Compact of Mayors. The Compact of Mayors is therefore a 'network of city networks'. The Compact of Mayors offers a form of certification of a city's climate performance. In order to demonstrate compliance with the Compact of Mayors, a city has to fulfil a series of steps that culminate in the public reporting of its GHG inventory, climate risks, and an action plan that provides details on how the city plans to achieve its GHG reduction target and improve its climate resilience. Regular updating of its GHG inventory, monitoring, and reporting is a compliance requirement. In the document *Compact of Mayors: Definition of Compliance*, it is stated, 'A complete updated inventory shall be required every four years, and the inventory year may be no more than four years prior to the reporting year.'[43] In addition, '[i]n between years when inventories are updated, "off-year reporting", cities shall report a list of: (1) improvements made to the quality of their inventory, focusing both on data availability and data quality; and (2) areas where outstanding data challenges exist'.[44] Upon fulfilling these requirements, a city will be certified 'compliant' and given a logo that it can use on its publicity materials. Like all voluntary schemes, the Compact of Mayors does not have any powers to force a city to comply except to withhold certification when a city does not meet the compliance requirements. It relies on reputational pressure and the perceived benefits of climate information disclosure to motivate cities to participate.

The GHG City Protocol was briefly discussed in Chapter 5 as a product of cooperation between C40, ICLEI, and other actors, including World Resources Institute, World Bank, and UNEP. The GHG City Protocol was developed because there was a need for a robust and widely applicable methodology that cities could use to calculate and report their GHG emissions. In other words, it is a compliance tool that advances the normative goals of urban climate law – i.e. that cities ought to pursue ambitious climate action because they are a major source of GHG emissions, and ambitious action ought to be undergirded by monitoring, reporting, and verification. This leads to the question of how regulatory tools such as the GHG

43 Compact of Mayors, 'Compact of Mayors: Definition of Compliance', online: www.bbhub.io/mayors/sites/14/2015/06/Compact-of-Mayors_Definition-of-compliance-082415.pdf (accessed on 1 July 2016).

44 Ibid.

City Protocol lead to behavioural change. That is, how does data collection and disclosure steer cities towards GHG emissions abatement and climate adaptation? While the pathways by which voluntary standards influence the behaviour of actors are varied and often context dependent, two pathways are of particular salience when it comes to the reduction of GHG emissions and pursuit of low-carbon development by cities. These pathways centre on the promotion of reflexivity amongst city officials and mayors, as well as norm diffusion.

6.3.1 *Promoting Reflexivity*

As compliance with voluntary standards is not mandated by law, advocates of these standards often use them to promote 'reflexivity' to bring about positive behavioural change. Gunther Teubner proposes reflexive law as a third and the latest stage in the evolution of legal systems.[45] Reflexive law 'seeks to design self-regulating social systems through norms of organization and procedure'.[46] Consequently, 'legal control of social action is indirect and abstract, for the legal system only determines the organizational premises of future action'.[47] At the heart of the concept of reflexivity lies the idea that 'social practices are constantly examined and reformed in light of incoming information about those very practices'.[48] Accordingly, the aim of laws that incorporate the reflexive concept is to create procedures and incentives that induce actors to assess their actions (hence the reflexivity) and adjust them to achieve socially desirable goals such as reducing GHG emissions, rather than dictating what to do in all cases. It is for this reason that in environmental law, for example, the emphasis on reflexivity has led to a notable trend of 'proceduralisation'.[49]

The use of environmental management systems (EMSs) is a prime example of reflexive regulation. An EMS may be described as 'a formal set of policies and procedures that define how an organization will manage its potential impacts on the natural environment and on the health and welfare of the people who depend on it'.[50] Organizations with an EMS typically adopt a written environmental policy; identify aspects of their activities, products, and services that have a deleterious impact on the environment; set goals to improve their environmental performance;

45 Gunther Teubner, 'Substantive and Reflexive Elements in Modern Law' (1983) 17 *Law and Society Review* 253. Eric Orts applied this concept to regulation; Eric W. Orts, 'Reflexive Environmental Law' (1995) 89 *Northwestern Law Review* 1227.

46 Gunther Teubner, ibid., pgs. 254–255. 47 Ibid.

48 Anthony Giddens, *The Consequences of Modernity* (Polity Press 1990), pg. 38. A related concept is reactivity; the idea underlying this concept is that 'individuals alter their behavior in reaction to being evaluated, observed or measured'; Wendy Nelson Espeland and Michael Sauder, 'Rankings and Reactivity: How Public Measures Recreate Social Worlds' (2007) 113 *American Journal of Sociology* 1.

49 Julia Black, 'Proceduralising Regulation: Part I' (2000) 20 *Oxford Journal of Legal Studies* 597.

50 Richard Andrews et al., 'Environmental Management Systems: History, Theory, and Implementation Research' in C. Coglianese and Jennifer Nash (eds), *Regulating from the Inside: Can Environmental Management Systems Achieve Policy Goals* (Resources for the Future 2001), pg. 32.

assign responsibility for implementing the initiatives to meet the targets; and have a process for evaluating and refining the EMS for further improvement in the future. In this manner, an EMS puts key processes in place to foster iterative learning.[51] There are several EMS models, but the most influential one by far is ISO 14001, developed by the International Organization for Standardization (ISO).[52] EMSs that require third-party certification such as the ISO 14001 standard and the European Union Eco-Management and Audit Scheme (EMAS) are routinely used, particularly in the automobile industry.[53] Research has shown that firms adopt EMSs for reasons including regulatory pressure,[54] enhancement of their corporate image,[55] and stewardship motivated by environmental values and community relationships.[56] Reviews of EMS programmes from various industries have found mixed evidence of improved environmental performance. For example, Ziegler and Rennings found that EMS certification did not significantly affect environmental innovation and pollution abatement behaviour at German manufacturing facilities.[57] Based on Japanese facility-level data, Arimura et al. find that ISO 14001 implementation was effective in reducing natural resource use, solid waste generation, and water wastage.[58] A study of the Finnish pulp and paper industry concluded that the greatest positive impact of EMSs arises when the system identifies previously unknown areas of environmental improvement for the firm.[59] It appears to be the case that EMSs can promote effective learning that leads to incremental improvement, but it is not likely to give rise to major innovations that require substantial investment.[60] It can be argued that EMSs should therefore be part of a suite of regulatory tools and that it is insufficient to rely solely on EMSs to manage environmental externalities. Used alongside traditional enforcement mechanisms, EMSs can lead to improved environmental performance over time.

Many of the voluntary standards and compliance tools that aim to help cities reduce their GHG emissions are variations of the EMS. To take part in the Compact

[51] The US EPA describes the system as a 'repeating cycle' that allows continuous improvement to occur; online: www.epa.gov/ems/learn-about-environmental-management-systems#what-is-an-EMS (accessed on 10 January 2016).

[52] Daniel J. Fiorino, *The New Environmental Regulation* (MIT Press 2006), pg. 102.

[53] Toshi H. Arimura, Akira Hibiki, and Hajime Katayama, 'Is a Voluntary Approach an Effective Environmental Policy Instrument? A Case for Environmental Management Systems' (2008) 55 *Journal of Environmental Economics and Management* 281, pg. 282.

[54] Nicole Darnall, Why US Firms Certify to ISO14001: An Institutional and Resource-Based View (Best Paper Proceedings of the 2003 Academy of Management Conference, Seattle, WA, 2003).

[55] Andreas Ziegler and Klaus Rennings, Determinants of Environmental Innovations in Germany: Do Organizational Measures Matter? A Discrete Choice Analysis at the Firm Level (ZEW Discussion Paper No. 04–30, Mannheim, 2004).

[56] John Cary and Anna Roberts, 'The Limitations of Environmental Management Systems in Australian Agriculture' (2011) 92 *Journal of Environmental Management* 878, pg. 881.

[57] Ziegler and Rennings. [58] Arimura, Hibiki, and Katayama.

[59] Mikeal Hilden et al., *Evaluation of Environmental Policy Instruments: A Case Study of the Finnish Pulp and Paper and Chemical Industries* (Helsinki: Finnish Environmental Institute 2002), pg. 113.

[60] Ibid.

of Mayors, city government officials are committed to undertaking a multi-year iterative process that will unlock the information required for self-reflection and identify weaknesses and previously unknown opportunities for GHG emissions abatement.[61] In other words, compliance with the Compact of Mayors is about following a guided process of data collection and assessment that will induce city governments to assess their existing policies and programmes, and adjust them to achieve GHG reduction and climate resilience targets. Cities, through their networks, have also developed the accounting and reporting tools they need to promote reflexivity. Global online reporting platforms like carbon*n* Climate Registry allow city government officials to easily peruse the vast database of city initiatives. They are able to track how their counterparts are performing. Benchmarking against the performance of other cities can be an important component of reflexive learning.[62] The use of a common set of standards to account and report citywide GHG emissions is crucial for promoting reflexivity, because benchmarking and tracking can be done effectively only when the data reported by cities is comparable. Hence, the promulgation of the GHG City Protocol is significant in terms of strengthening reflexivity as a pathway of influence by which cities steer their behaviour towards attaining climate mitigation and adaptation objectives. Finally, there are online GHG inventory tools such as *ClearPath* that have been developed by cities, through the network ICLEI, and are made available free of charge to city governments.[63] This helps overcome the cost barriers to effective data collection, which is expensive. For example, monitoring the implementation of the UN's Sustainable Development Goals is estimated to require US$1 billion in aid assistance to support data collection in developing countries alone.[64]

6.3.2 *Norm Diffusion*

Theories of diffusion offer an important account of how norms and voluntary standards eventually become widely adopted by a community. According to these theories, 'norm entrepreneurs' – such as international organizations, transnational advocacy networks, and epistemic communities – play an important role in persuading a 'critical mass' of relevant actors to embrace a new norm and become 'norm followers'. When a 'tipping point' is reached, we can expect the norm to 'cascade' through the rest of the community and become widely adopted within a fairly circumscribed period of time.[65] Diffusion occurs in the absence of formal or contractual obligations. In other words, the actor in question (e.g. a state or city)

61 Interview No. 6. 62 Interview Nos. 4 and 6.

63 *ClearPath* complies with the GHG City Protocol, and a city can use it to report directly to carbon*n* Climate Registry; online: www.clearpath.global/features/ (accessed on 12 July 2016).

64 Jessica Espey et al., *Data for Development: A Needs Assessment for SDG Monitoring and Statistical Capacity Development* (Sustainable Development Solutions Network 2015), online: http://unsdsn.org (accessed on 1 July 2016).

65 Finnemore and Sikkink, pg. 895.

does not have any formal commitments towards a government or an international organization to implement the norm in question, unlike cases of harmonization or coercion in international politics.[66] Briefly, 'harmonization' refers to the conscious modification of policies by governments committed to transnational standards that they have had a hand in crafting.[67] States are primarily motivated to engage in processes of international harmonization when the problem that needs to be solved cannot be addressed without collective action.[68] Another motivation for harmonization is to reduce barriers to the free movement of goods, capital, and people.[69] Coercion occurs when states, international organizations, or private actors use asymmetrical power relationships to dictate policies to others.[70] Economic or political conditionality are more common forms of coercion than the use of military force. The EU, for example, influences the domestic policies of Central and Eastern European countries by linking the opening of the accession negotiations to political reforms aimed at improving the protection of human rights.[71] In contrast to harmonization and coercion, diffusion is a process whereby policymakers make independent decisions but do take into account policy choices in other jurisdictions.[72]

There are myriad diffusion mechanisms, but most mechanisms can be grouped into three broad categories: competition, learning, and imitation.[73] Briefly, competition would involve cities influencing one another's policy choices as they compete to attract economic resources. Learning occurs when policymakers look across jurisdictional borders towards other cities in search of effective solutions to domestic problems. Imitation takes place when a norm is adopted because its socially constructed meaning matters more than its objective consequences. It should be noted that these categorizations are adopted as an analytical schematic, but they are by no means mutually exclusive. City A may choose to adopt a voluntary standard because it wants to be perceived as being a responsible global citizen (imitation). The city's

[66] Helge Jörgens, 'Governance by Diffusion: Implementing Global Norms through Cross-National Imitation and Learning' in William M. Lafferty (ed), *Governance for Sustainable Development: The Challenge of Adapting Form to Function* (Edward Elgar 2004), pg. 252.

[67] Michael Howlett, 'Beyond Legalism? Policy Ideas, Implementation Styles and Emulation-Based Convergence in Canadian and U.S. Environmental Policy' (2000) 20 *Journal of Public Policy* 305, pg. 308.

[68] Jörgens, pg. 251. [69] Ibid.

[70] Farizio Gilardi, 'Transnational Diffusion: Norms, Ideas and Policies' in Walter Carlsnaes, Thomas Risse, and Beth A. Simmons (eds), *Handbook of International Relations* (2nd edn, Sage Publishing 2012), pg. 461.

[71] Ibid. Also see Frank Schimmelfenniga and Ulrich Sedelmeierb, 'Governance by Conditionality: EU Rule Transfer to the Candidate Countries of Central and Eastern Europe' (2004) 11 *Journal of European Public Policy* 661.

[72] Elkins and Simmons, pg. 35. Gilardi emphasizes that diffusion is not the same as convergence: 'diffusion is the process that leads to the pattern of [policy] adoption … [while] convergence characterizes the outcome of the process'; Gilardi, pg. 455.

[73] A diffusion mechanism is 'a systematic set of statements that provide a plausible account of why the behavior of A influences that of B'; Dietmar Braun and Farizio Gilardi, 'Taking 'Galton's Problem' Seriously: Towards a Theory of Policy Diffusion' (2006) 18 *Journal of Theoretical Politics* 298, pg. 299.

policymakers may also have evaluated case studies and concluded that adopting City B's practices is likely to deliver significant benefits (learning). City A's policymakers may also take into account that implementation of a particular voluntary standard will enhance the city's 'brand' as an environmentally progressive and attractive place to live and work in, thus making the city more attractive to global talent.[74] While this may be described as a form of competitive behaviour, it is not competition as is typically understood in the literature.[75] In the present case, learning and imitation best explain how urban climate law has become widely adopted amongst cities.

6.3.2.1 Learning

Learning can be characterized as a process of rational and problem-oriented 'lesson drawing'. Policymakers focus on the merits and outputs of a certain policy, and their motivation for adopting the policy in question is its perceived capacity to improve regulation.[76] Let us take, for example, Chinese president Xi Jinping's announcement during his visit to Washington, DC, in September 2015 that China plans to implement a national emissions trading scheme (ETS) by 2017.[77] Preceding the development of a national scheme, the central government in Beijing authorized the establishment of pilot ETSs in seven provinces. The central government deliberately chose seven provinces that are at different levels of economic and industrial development so that emissions trading under different conditions can be tested and better understood.[78] The provincial authorities responsible for designing and implementing the pilot schemes were given considerable leeway to adopt different approaches to regulatory design issues, such as the choice of economic sectors included in the pilot scheme and the methods of allocating allowances to regulated entities.[79] At the same time, Chinese government officials and researchers looked closely at the experiences of other jurisdictions, notably the European Union Emissions Trading Scheme (EU ETS), which is the oldest and largest emissions trading scheme globally.[80]

74 Ida Andersson, '"Green Cities" Going Greener? Local Environmental Policy-Making and Place Branding in the "Greenest City in Europe"' (2016) 6 *European Planning Studies* 1197; Mihalis Kavaratzis and G. J. Ashworth, 'City Branding: An Effective Assertion of Identity or a Transitory Marketing Trick?' (2005) 96 *Tijdschrift voor economische en sociale geografie* 506.

75 See discussion in Section 6.3.2.3.

76 Richard Rose, 'What Is Lesson-Drawing?' (1991) 11 *Journal of Public Policy* 3.

77 Demetri Sevastopulo and Pilita Clark, 'Xi, Obama and the Pope Raise Climate Hopes' *Financial Times* (25 September 2015).

78 ZhongXiang Zhang, Carbon Emissions Trading in China: The Evolution from Pilots to a Nationwide Scheme [Centre for Climate Economic & Policy (Australian National University), CCEP Working Paper 1503, 2015], pg. 11, online: https://ccep.crawford.anu.edu.au/sites/default/files/events/attachments/2015–04/paper_by_professor_zhang.pdf (accessed on 1 July 2016). Also see Pilita Clark, 'The "Black Hole" of Chinese Carbon Trading' *Financial Times* (13 May 2014).

79 Zhang, ibid.

80 For detailed overview, see European Commission, *EU ETS Handbook*, online: http://ec.europa.eu/clima/publications/docs/ets_handbook_en.pdf (accessed on 1 July 2016).

It can be argued that when Chinese policymakers use the information gathered from the experience of the EU ETS and the pilot schemes to update their understanding of how emissions trading works in practice, their understanding will shift towards what the pilot experiments have demonstrated.[81] However, the extent of this shift in beliefs will depend on the consistency of the information received and the strength of the Chinese policymaker's prior convictions.[82] This leads to my next point that learning seldom occurs in a 'neat' and rational manner. Like everyone else, policymakers are bounded by cognitive limits or rely on multiple strategies to cope with cognitive constraints.[83] Herbert Simon, the economist behind the theory of bounded rationality, argued that individuals do not seek to maximize benefit from a particular course of action, as they are not capable of assimilating all the information necessary to do so. Even if they could gain access to all the necessary information, the human mind is not able to process the information properly, and therefore necessarily adopts certain 'short cuts' and restrictions.[84] Thus, when a city's planners engage in learning, it may simply be a case of adopting a certain standard because information about it is most easily accessible. They may also be more keen to adopt policies and practices from jurisdictions which they are more familiar with (e.g. similar cultural attributes such as the language its people speak and the country's colonial origins).[85] In this line of thinking, the Chinese chose to learn from the EU's experience not only because it was rational to do so, but also because information and access to EU ETS policymakers was readily available as a result of the EU's active climate diplomacy aimed at developing 'a global network of emissions trading systems'[86] that will function as a global carbon market.[87]

In the case of urban climate law, learning is an important norm diffusion mechanism that is facilitated by a number of factors. In Mexico City, for example, when city government officials explored a range of solutions to reduce traffic

[81] Gilardi, pg. 465.

[82] A good indication that 'lesson drawing' has occurred is that all the Chinese pilot schemes have one feature in common: they all incorporate mechanisms to address supply–demand fluctuations and resulting price volatility – a significant problem that has adversely affected the performance of the EU ETS.

[83] Fiske and Taylor famously characterized the social perceiver as a 'cognitive miser'; Susan T. Fiske and Shelley E. Taylor, *Social Cognition* (1st edn, Addison-Wesley Publishing 1984). In the 1991 revised edition of their classic text, they suggested that the cognitive miser metaphor ought to be replaced with one of the 'motivated tactician, a fully engaged thinker who has multiple cognitive strategies available and chooses among them based on goals, motives and needs'; Susan T. Fiske and Shelley E. Taylor, *Social Cognition* (McGraw-Hill 1991), pg. 13.

[84] It is widely believed that Simon first used the expression 'bounded rationality' in his 1957 work; Herbert Simon, *Models of Man* (Wiley & Sons 1957).

[85] Elkins and Simmons, pgs. 44–45.

[86] European Commission, *Questions and Answers on the revised EU Emissions Trading System* (2008), online: http://europa.eu/rapid/press-release_MEMO-08-796_en.htm?locale=en (accessed on 1 July 2016).

[87] For discussion, see Kati Kulovesi, Elisa Morgera, and Miquel Muñoz, 'Environmental Integration and the Multifaceted International Dimensions of EU Law: Unpacking the EU's 2009 Climate and Energy Package' (2011) 48 *Common Market Law Review* 829.

congestion and improve public transportation within the city to reduce GHG emissions and conventional air pollutants, the officials looked to other cities known to have successfully addressed the same problem, such as Bogotá in Colombia.[88] When it came to efforts to make Mexico City more bike friendly, they looked to Amsterdam in the Netherlands.[89] In other words, they tried to look for instances of best practices. In the past few years, the number of organizations that produce case studies of urban climate mitigation and adaptation best practices that are made available at no cost has grown exponentially.[90] For many city governments, the issue is not the lack of information about best practices, but rather how to evaluate and decide on the best practice that will be most suitable and effective for their city.[91] Initiatives such as C40's Connecting Delta Cities network also play an important role in facilitating the learning process as a means of diffusing urban climate law. Through the Connecting Delta Cities network, Rotterdam's sustainability advisors worked closely with their counterparts in Ho Chi Minh City to develop the Vietnamese city's resilience strategy for managing flood risks.[92] Through this cooperative process, the norms, practices, methodologies, and voluntary standards that constitute urban climate law were diffused from one city (Rotterdam) to another (Ho Chi Minh City) across continents.

Official aid and technical assistance from national governments also play a role in providing more information about certain cities over others, and therefore influence the learning process. For example, over the years, the governments of the Netherlands, the United Kingdom, and Japan have provided Mexico with technical assistance for environmental protection initiatives.[93] Through capacity-building programmes, Mexico City officials became acquainted with examples from cities in these countries and are therefore more likely to learn from these cities.[94] Learning can also be facilitated by official bilateral cooperation between cities. For example, Singapore's Centre for Liveable Cities has hosted government officials from Chinese cities, who spent up to three months in Singapore learning about its urban governance strategies, with the goal of ascertaining how these strategies can be transplanted to cities in China.[95]

Non-profit organizations also influence how learning proceeds as a norm diffusion mechanism. The Hewlett Foundation, for example, is an American private

[88] Interview No. 6. [89] Ibid.

[90] Junichi Fujino et al., *City Champions: Scaling-Up Transformative Sustainability Innovations* (IGES Discussion Paper prepared for the International Forum for Sustainable Asia and the Pacific 2016), pg. 4, online: http://pub.iges.or.jp/modules/envirolib/upload/6671/attach/CTF_ISAP_Paper_070716_final.pdf (accessed on 1 July 2016).

[91] Ibid.

[92] C40, 'With Help from Rotterdam, Ho Chi Minh City Launches Climate Adaptation Strategy', online: www.c40.org/blog_posts/with-help-from-rotterdam-ho-chi-minh-city-launches-climate-adaptation-strategy (accessed on 1 July 2016).

[93] Interview No. 6. [94] Ibid.

[95] Centre for Liveable Cities (Singapore), 'Shanghai Officials Share Thoughts on Singapore Stint', online: www.clc.gov.sg/Training/trainingprogrammes.htm (accessed on 1 July 2016).

foundation that awarded approximately US$400 million of grants in 2015 to solve environmental and social problems.[96] It has actively supported civil society organizations and the government of Mexico City by providing grants, carrying out training programmes, and lending expertise to help the city tackle its air pollution problem and mitigate climate change.[97] In this manner, Hewlett Foundation has introduced many best practices from major American cities such as Los Angeles and San Francisco to Mexico City's public officials, who have drawn lessons and incorporated some of these practices and policies in their domestic strategies.[98]

Another example is the Rockefeller Foundation. The foundation established the 100 Resilient Cities programme, which promotes a holistic approach towards tackling climate change by addressing it as part of the numerous, interconnected physical, social, and economic challenges that cities face.[99] It promotes the idea of resilience, which includes adaptation to the risks of climate change, reduction of social inequality, and a society's cybersecurity practices (cyber resilience).[100] When a city becomes a member of the 100 Resilient Cities programme, the Rockefeller Foundation provides it a grant which pays the salary of the city's chief resilience officer, who has overall responsibility for spearheading and coordinating the city's resilience initiatives.[101] The city of Rotterdam is one of the first cities to join the 100 Resilient Cities programme.[102] Today, it has a progressive and ambitious resilience strategy that involves numerous stakeholders, including the city's social welfare agency, the environment and water resources department, schools, museums, small- to mid-sized businesses, and clean technology consortiums. According to one of the city's sustainability advisors who has worked on climate adaptation and subsequently resilience since the mid 2000s, joining 100 Resilient Cities proved to be a turning point for Rotterdam's climate change strategy.[103] Prior to joining the programme, Rotterdam's focus was on adaptation to flooding and other water-related risks. However, 'joining 100 Resilient Cities was very significant in terms of changing [city government officials'] attitudes from focusing on water adaptation to thinking more broadly about how we can design public spaces to be climate-proof. This thinking helped connect climate adaptation to spatial design in Rotterdam. Resilience gave us a new lens to look at connecting policies and integrating policy fields in this city.'[104] To gain familiarity with the concept of resilience, Rotterdam's city officials found the 100 Resilient Cities reading materials, research toolkits (such as the City Resilience Framework[105]), and workshops very helpful for overcoming the initial learning curve and triggering new thinking about the application of the resilience concept in Rotterdam.[106]

96 Hewlett Foundation, online: http://hewlett.org/about-us (accessed on 1 July 2016).
97 Interview Nos. 6 and 10. 98 Ibid.
99 100 Resilient Cities, online: www.100resilientcities.org/#/-_/ (accessed on 1 July 2016). 100 Ibid.
101 Ibid.
102 100 Resilient Cities, 'First Resilient Cities Announced by Rockefeller Foundation', online: www.100resilientcities.org/blog/entry/33-resilient-cities-announced#/-_/ (accessed on 1 July 2016).
103 Interview No. 11. 104 Ibid. 105 100 Resilient Cities. 106 Interview No. 11.

6.3.2.2 Imitation

Imitation occurs when a mayor emulates a policy or adheres to a norm to maximize his/her city's reputation or his/her personal reputation. In this line of thinking, norms may be understood as common practices whose value to an actor stems from their prevalence in a community.[107] The predominant benefit of norm adherence is reputational; adhering to the norm and thereby joining a growing majority of actors confers a degree of legitimacy upon the city and its mayor.[108]

For example, after a personal visit by representatives of the Clinton Climate Initiative and C40, the potential reputational gain was a key factor that convinced the mayor of Rotterdam to start taking concerted action on climate change.[109] The mayor saw the strategic value for Rotterdam, a relatively small global city, to become a member of C40. Rotterdam could then project the image of being a global city that is in the same league as New York City and London when it comes to addressing climate change. According to the world polity school, states enter into international human rights treaties to signal their adherence to global cultural norms that are perceived to be 'universal', 'advanced', and 'modern'.[110] In a similar manner, urban climate law performs an expressive function when global cities sign up to the Compact of Mayors. Committing to the Compact signals a city's adherence to the norms underpinning ambitious climate action that are perceived to be 'progressive' and 'global'. The cumulative effect of many cities signing up for the Compact of Mayors is analogous to 'peer pressure' amongst cities, and three possible motivations for responding to such peer pressure are the desire for legitimacy, conformity, and the quest for esteem.[111]

An alternative (and not mutually exclusive) explanation is premised on a less instrumentalist attitude towards norms. According to such an explanation, a mayor chooses to adhere to a particular set of norms in her policies because she believes in the appropriateness of these norms. The logic of appropriateness is such that '[a]ction involves evoking an identity or role to a specific situation'.[112] From a constructivist perspective, actors can genuinely change their understanding of what constitutes appropriate behaviour through discursive interaction. Such interaction is said to generate knowledge, particularly so-called rhetorical knowledge, which refers to knowledge that is offered or created in dialogue and employed in practical reasoning.[113] C40 and other organizations that seek to promote particular

[107] Elkins and Simmons, pg. 39. [108] Ibid. [109] Interview No. 11.

[110] Didem Buhari-Gulmez, World Polity School (Oxford Bibliographies 2014), online: www.oxfordbibliographies.com/view/document/obo-9780199743292/obo-9780199743292–0019.xml (accessed on 1 July 2016).

[111] Finnemore and Sikkink, pg. 903.

[112] James G. March and Johan P. Olsen, 'The Institutional Dynamics of International Political Orders' (1998) 52 *International Organization* 943, pg. 951.

[113] See discussion in Chapter 3; on rhetorical knowledge, see Francis J. Mootz III, 'Natural Law and the Cultivation of Legal Rhetoric' in Willem J. Witteveen and Wibren Van der Burg (eds), *Rediscovering Fuller: Essays on Implicit Law and Institutional Design* (Amsterdam University Press 1999), pgs. 442–448.

aspects of urban climate law recognize the importance of discursive interaction and the role it plays in shaping global cities' understanding of those norms and practices. C40, for example, regularly organizes workshops and web-based seminars (webinars) to create opportunities for city officials to interact and generate ideas.[114]

Another example is Matchmaker, a new initiative of the *Low Carbon City Lab*, which is a public-private partnership that aims to leverage 25 billion euros' worth of climate finance for cities by 2050.[115] Matchmaker aims to be a platform that introduces investors and city administrations to each other, with the objective of helping cities secure funding for their climate mitigation and adaptation projects.[116] As early as its first year of operations, Matchmaker convened a number of workshops around the world to bring representatives of cities, development banks, sovereign funds, commercial banks, and civil society together to discuss and share their views about emerging climate finance practices and standards.[117] These workshops are convened based on the understanding that norm adoption and adherence is much more likely when the actor in question believes in the appropriateness of the norm.[118]

Thus, within the transnational climate change regime complex where the adoption of urban climate law is voluntary, various actors have created forums for engagement and discursive interaction amongst cities and other stakeholders. These forums play an important role in facilitating the diffusion of urban climate law by learning and imitation amongst cities.

6.3.2.3 Competition

In international relations, regulatory competition is said to occur when states struggle to shape policy developments at the international level to accord with their national policy interests and legal traditions. In this line of thinking, the state (or group of states) that manages to significantly shape a regulatory policy according to its preferences will yield the most benefits, as that state will have to undertake the least amount of political and economic adjustments in response to the new rule. Conversely, the states that have to make the biggest political and economic adjustments to the new rule are deemed to be at a disadvantage.[119] Norm diffusion occurs when a state (or group of states) manages to successfully assert its regulatory influence such that its proposed rule or norm becomes widely adopted in the international community.

114 Interviews Nos. 5, 6, 8, 10, and 11.

115 Low Carbon City Lab, online: http://local.climate-kic.org/ (accessed on 1 July 2016). 116 Ibid.

117 Interview No. 2.

118 This is a conclusion I drew from my observations of a Matchmaker workshop in progress (Bonn, 5 July 2016).

119 K. Kern, H. Jörgens, and M. Jänicke, *The Diffusion of Environmental Policy Innovations: A Contribution to the Globalisation of Environmental Policy* [WZB discussion paper FS II 01–302, Berlin: Social Science Research Center (WZB), 2001], pgs. 4–5; K. Tews, P. Busch, and H. Jörgen, 'The Diffusion of New Environmental Policy Instruments' (2003) 42 *European Journal of Political Research* 569.

Another variety of regulatory competition can be defined as a process whereby rules are selected and de-selected through competition between decentralized, rule-making entities, which can be states or other political units such as regions.[120] The competitive process is expected to yield a number of beneficial outcomes. First, it promotes diversity and experimentation amongst the competing jurisdictions in the search for effective laws.[121] Secondly, when users of the rules are able to express their preference for some rules (and not others), it promotes the flow of information on what works better in practice.[122] Finally, it can be argued that regulatory competition leads to the content of rules becoming more effectively tailored to meet the needs of users.[123] In the model set out in Charles Tiebout's seminal article 'A Pure Theory of Local Expenditures', local governments compete to attract residents by offering different packages of facilities and public services that are made available at different tax rates.[124] The resident is assumed to pick the community that offers the public goods and services that best suits his or her preferences. The end result is to maximize the welfare of residents as well as maintain diversity amongst jurisdictions. While Tiebout's model was concerned with the provision of public goods and services by local governments, laws can also be viewed as a type of public good. The diffusion of norms occurs when, for example, jurisdictions learn from one another, seek to export their norms and practices, and put selected norms into practice.

It can be argued that both variants of regulatory competition described here do not explain how urban climate law is diffused across global cities. There is scant evidence of global cities actively seeking to influence the normative content of urban climate law at the transnational level in order to secure first-mover advantage. There is also little evidence that global cities are seeking to offer different versions of urban climate law in order to attract residents and businesses. Certainly, there is diversity in the practices and regulations to address climate change in global cities across the world. However, this diversity is attributable to the need for global cities to localize and adapt urban climate law to suit domestic needs and circumstances rather than the dynamics of regulatory competition.

6.3.3 *The Role of Norm Entrepreneurs*

Norm entrepreneurs play an important role in promoting norms and strategically providing information on the range of policy tools available. Transnational channels of communication provide norm entrepreneurs with the basic platform to

120 Simon Deakin, 'Is Regulatory Competition the Future for European Integration?' (2006) 13 *Swedish Economic Policy Review* 71, pg. 74.

121 Simon Deakin, Legal Diversity and Regulatory Competition: Which Model for Europe? (Working Paper No. 323, Cambridge: Centre for Business Research, University of Cambridge 2006), pg. 4.

122 Ibid. 123 Ibid.

124 Charles M. Tiebout, 'A Pure Theory of Local Expenditures' (1956) 64(5) *Journal of Political Economy* 416.

disseminate information about new policy instruments, institutions, and best practices. The information about best practices is usually presented in a sufficiently abstract manner so that it can be adapted to meet the needs of different politico-institutional settings.[125] One of the important things that norm entrepreneurs do is raise awareness of the problem or the 'wrong' that new norms or standards are intended to address.

It is not difficult to see how mayors like Marcelo Ebrard of Mexico City and Ken Livingstone of London acted as important norm entrepreneurs when they made tackling climate change a focus of their mayoralty.[126] As chairperson of the World Mayors Council on Climate Change and the mayor of the first city in the LAC region to adopt GHG reduction targets and a climate strategy, Marcelo Ebrard was in a strong position to persuade other cities, particularly those in the LAC region, to adopt some of the policies that his city had implemented and the norms underlying urban climate action. Ken Livingstone's office developed an energy strategy that committed London to reducing carbon dioxide emissions by 20 per cent below 1990 levels by 2010 as a first step to a reduction of 60 per cent by 2050. The mayor's office also publicly backed high-profile projects that demonstrated the technical feasibility of renewable energies, in addition to developing multi-stakeholder partnerships for research and development of new hydrogen technologies. During this period, climate change was high on the global political agenda, and there was a great deal of public concern about the stalemate in international negotiations. By spearheading the development of London's first climate change strategy, Livingstone sought to demonstrate the potential that cities have to address climate change and to develop a new normative narrative that cities ought to address climate change because cities are a main contributing source to climate change but are also capable of finding innovative solutions. By founding what is now the C40 network, Livingstone created an important transnational mechanism for the diffusion of the norms and practices that lead to urban climate mitigation and adaptation.

Transnational city networks like C40 also act as important norm entrepreneurs. As discussed earlier as well as in Chapter 5, C40 has set up networks to make information on best practices available to its members to encourage learning. It further facilitates learning by supporting these issue-specific networks in organizing webinars and workshops to disseminate ideas and provide forums for city officials to learn from each other. C40's research team also generates original research, policy recommendations, and case studies. It can also be argued that C40 fosters imitation as a norm diffusion mechanism by providing information on the individual performance of cities. More specifically, C40 does this by making the adoption of a climate strategy and regular reporting key components of its benchmarking system. City members can compare their performance results via the CDP website, the C40

[125] Lukas Hakelberg, 'Governance by Diffusion: Transnational Municipal Networks and the Spread of Local Climate Strategies in Europe' (2014) 14 *Global Environmental Politics* 107, pg. 114.

[126] The information in the rest of this paragraph is drawn from Chapter 4.

website, and the Climate Action in Megacities survey. Through the practice of benchmarking, C40 seeks to enhance the perception that adopting a local climate strategy is appropriate and commendable and to foster a healthy dose of competition amongst the cities to improve their performance. Further, by making information about the performance of city members easily accessible, C40 enables its members and interested parties to identify laggard cities and use a combination of carrots and sticks to prompt these cities to undertake more mitigation and adaptation actions.

The foregoing discussion considered how cities, through their networks, create urban climate law, which includes norms, practices, and voluntary standards such as the GHG Protocol for Cities and the Compact of Mayors. These standards lead to the reduction of urban GHG emissions through two key mechanisms: (1) promoting reflexivity and (2) norm diffusion processes. Through these pathways of influence, the norms created by cities and transmitted via transnational city networks are changing the behaviour of cities that have elected to adopt them. While cities are unable to create mandatory rules at the international level, cities are undoubtedly engaged in transnational voluntary rule-making. Urban climate law also interacts with the UNFCCC rules and institutions. The next section argues that the interaction between urban climate law and the UNFCCC regime has had mutually reinforcing effects and has resulted in an overall expansion of governance within the transnational climate change regime complex. There is, accordingly, no conflict between urban climate law and the inter-state UNFCCC regime. On the contrary, we see both types of authority linked in strategic and deliberate ways.

6.4 THE INTERACTION BETWEEN URBAN CLIMATE LAW AND THE UNFCCC REGIME

The emergence of urban climate law can be viewed as part of a constellation of 'governance experiments' that have emerged in response to the inadequacies of the UNFCCC treaty regime.[127] Cities have therefore had the opportunity to survey the problems with the UNFCCC's response to climate change and the issues that the treaty regime has not been able to address. It can be argued that cities have then sought to take a different tack regarding the climate change problem and develop alternative solutions to address aspects of climate change that the UNFCCC regime has struggled to. At the same time, cities have not been apathetic to the role of the UNFCCC in governing climate change and, in fact, lobby for more robust and ambitious action by states.[128] This has led to coupling, whereby urban climate law has been deliberately designed to complement and strengthen the UNFCCC regime.

[127] See discussion in Chapter 2.

[128] ICLEI, 'Local Government Climate Roadmap', online: www.iclei.org/index.php?id=1197 (accessed on 1 July 2016).

I argue that cities have taken a different approach towards the climate change problem by *reframing* the issue and developing an alternative set of rules and practices to those of the UNFCCC regime. The voluntary standards created by cities and their networks affect the evolution of the transnational climate change regime complex as well as reinforce the UNFCCC regime by serving as a means to *diffuse the authority of the UNFCCC regime*. In this view, cities serve as additional venues for the use and adoption of the UNFCCC norms.[129] I will illustrate this point later with reference to C40's adherence to the CBDRRC principle, which is a cornerstone of the UNFCCC, the Kyoto Protocol, and the Paris Agreement. C40's adherence to the principle is de facto, not de jure, as it is not legally bound by the international climate change treaties. By voluntarily adopting this principle, and applying it within the network, C40 is indirectly expanding the authority of the UNFCCC regime. The Compact of Mayors serves as the third illustrative example of coupling. Cities have made a concerted effort to ensure that city-level governance efforts complement international efforts to develop MRV practices such that urban climate law and the UNFCCC regime mutually reinforce and support the development of an emerging norm of transnational climate law.

6.4.1 Reframing the Problem

The early stages of climate governance were guided by a dominant definition of the problem shared by the international community – that climate change is a collective action problem on a global scale and therefore required multilateral treaty-making based on negotiations that involved all states.[130] This thinking was reinforced by influential ideas such as the 'matching principle' proposed by Butler and Macey, which holds that the level of jurisdictional authority should match the scale of the harm being regulated, to ensure that all of the costs of the activity are internalized within the jurisdiction so as to prevent free-riding.[131] In relation to global environmental problems like climate change, the 'matching principle' calls for international regulation. Adler, following Butler and Macey, argues that climate change presents an unequivocal case for action at the national and international levels.[132] Accordingly, cities had little, if any, contribution to make towards addressing the problem. A conflux of factors subsequently created a 'policy window' for cities to

[129] This line of argument was inspired by the discussion on how private authority may affect the evolution of a regime complex in Jessica Green and Graeme Auld, 'Unbundling the Regime Complex: The Effects of Private Authority' (2016) *Transnational Environmental Law*, DOI: http://dx.doi.org/10.1017/S2047102516000121 (published online: 20 May 2016).

[130] See discussion in Chapter 2.

[131] Jonathan R. Macey and Henry N. Butler, 'Externalities and the Matching Principle: The Case for Reallocating Environmental Regulatory Authority ' (1996) 14 *Yale Law and Policy Review* 23, pg. 25.

[132] Jonathan H. Adler, 'Jurisdictional Mismatch in Environmental Federalism' (2005) 14 *New York University Environmental Law Journal* 130, pg. 175–176.

assert themselves on the governance landscape and successfully project authority.[133] As discussed in Chapter 3, amongst these factors were (1) the efforts by the World Bank, UN-Habitat, UNESCO, and various other international organizations to enlist cities as 'partners' to address a host of governance challenges including racism, gender discrimination, and public health in city slum settlements and (2) the global trend of political decentralization and devolution. As responsibilities for providing services to citizens have been passed on from national governments to provincial governments and on to city authorities, cities have experienced strain on their resources but at the same time are emboldened in their claims to authority.

Taking advantage of the policy window, cities reframed the climate change problem in ways that allowed them to carve out a meaningful role for themselves. First, cities have asserted that climate change requires a multilevel governance solution, and that includes the international level as well as the local level. In this narrative, the fact that cities account for 37–49 per cent of global GHG emissions and urban infrastructure accounts for over 70 per cent of global energy use is of notable significance.[134] With this sizable contribution by cities to global GHG emissions, it follows that cities play a critical role in global climate policy, and it is in cities 'where the struggle to mitigate climate change will be either won or lost'.[135]

Secondly, cities have sought to redirect the focus of climate policy from emission reductions to understanding emission sources. The hope is that this understanding, gained from the use of accounting and reporting standards, will promote reflexivity and policy diffusion leading to future reductions in GHG emissions. Norm entrepreneurs like Michael Bloomberg and CDP (the official platform for cities to report their climate actions) have keenly reiterated that the ultimate goal of climate action is redirecting our economies and societies onto a low-carbon pathway. In this regard – and this is the third aspect of the reframing – cities have positioned themselves as well placed to lead the economic and societal transformations. In particular, cities can serve as 'laboratories of democracy' to test how different policy options for decarbonization work in practice as well as demonstrate to state, federal, and international policymakers that climate action is not only possible but can also be cost-effective.[136] Fourthly, like many other transnational climate governance 'experiments', cities and their networks emphasize the economic gains from tackling

[133] On the concept of a policy window, see John Kingdon, *Agendas, Alternatives and Public Policies* (Longman 1995).

[134] Leadership, School, and ICLEI, pg. 9.

[135] UN-Habitat, 'UN-Habitat launched Guiding Principles for City Climate Action Planning at the Climate Change Conference (COP21) in Paris', online: http://unhabitat.org/un-habitat-launches-guiding-principles-for-city-climate-action-planning-at-the-climate-change-conference-cop21-in-paris/ (accessed on 1 July 2016).

[136] The term 'laboratories of democracy' was coined by Justice Louis Brandeis to depict the American states as testing grounds for innovative policies that can be replicated by other states or translated to the federal level; *New State Ice Co. v. Liebmann* (1932) United States Supreme Court No. 463. I have co-opted this term to refer to cities as policy testing grounds in the context of transnational climate change governance.

climate change.[137] They attempt to shift attention from the costs of reducing emissions, which has been a major bone of contention between developed and developing countries in the international climate change treaty negotiations, to the benefits that can be gained from reducing GHG emissions and making the transition towards low-carbon development. These benefits include economic diversification and job creation through developing the clean technology sector of the economy.

Finally, cities have reframed the climate change issue as one predominantly about reducing GHG emissions to one about the broader quest for sustainable urban development – that is, maximizing the benefits of urbanization while minimizing its ills as well as implementing the 2030 Agenda for Sustainable Development.[138] UN-Habitat has played an important role in this reframing as it seeks to address climate change through its specific lenses of managing urban growth and, through various projects and programmes, has been encouraging and empowering cities to address climate change as part of a broader agenda of sustainable urbanization.[139] These initiatives include the Cities and Climate Change Initiative, which '[builds] on UN-Habitat's long experience in sustainable urban development … [to help] counterparts develop and implement pro-poor and innovative climate change policies and strategies',[140] and the Urban Low Emissions Development Strategies project, which provides cities with technical assistance 'to integrate low-carbon strategies into all sectors of urban planning and development'.[141]

By reframing the issue, urban climate law has purposefully sidestepped contentious issues that have obstructed the UNFCCC regime, such as the costliness of reducing GHG emissions and the imposition of legally binding targets. In this way, urban climate law complements the UNFCCC regime by providing alternative and less-controversial ways to achieve GHG emission reductions.

6.4.2 *Urban Climate Law as a Means of Diffusing UNFCCC Norms*

The CBDRRC principle lies at the heart of the international climate change regime. Since the start of international discussions, this principle has underpinned the international community's efforts to address climate change. At the Second World Climate Conference in 1990, countries declared that the 'principle of equity and common but differentiated responsibility of countries should be

137 Matthew Hoffmann, *Climate Governance at the Crossroads: Experimenting with a Global Response after Kyoto* (Oxford University Press 2011), pg. 39.

138 UN General Assembly, *Transforming Our World: The 2030 Agenda for Sustainable Development* (Resolution 70/1, adopted by the General Assembly on 25 September 2015).

139 See discussion in Chapter 3.

140 UN-Habitat, 'Cities and Climate Change Initiative', online: http://unhabitat.org/urban-initiatives/initiatives-programmes/cities-and-climate-change-initiative/ (accessed on 21 July 2016).

141 UN-Habitat, 'Urban Low Emission Development Strategies', online: http://unhabitat.org/urban-initiatives/initiatives-programmes/urban-low-emission-development-strategies/ (accessed on 21 July 2016).

the basis of any global response to climate change'.[142] The CBDRRC principle is articulated in Article 3 of the UNFCCC and highlighted in the Kyoto Protocol and numerous UNFCCC COP decisions, including the Bali Action Plan of 2007 and the Cancun Agreements of 2010. Several strands of thought are brought together within this principle. First, the principle 'establishes unequivocally the common responsibility of [s]tates to protect the global environment'.[143] Secondly, the notion of differentiated responsibility is derived from the differences in the level of economic development and capabilities amongst states, as well as a country's historical contribution to climate change as a measure of its responsibility.[144]

The distinction between developed and developing countries has been captured in the rigid form of an annex to the UNFCCC; in UNFCCC parlance, 'Annex I states' refers to developed countries and countries in transition, and 'non-Annex I states' refers to developing countries. Based on the CBDRRC principle, the international climate regime carved out a leadership role for developed countries.[145] The UNFCCC states that Annex I parties should provide new and additional financial resources to non-Annex I parties. Further, they ought to facilitate and finance the transfer of environmentally sound technologies to non-Annex I parties. The UNFCCC goes on to state that the implementation of the Convention by developing countries will depend on the effective implementation by developed countries of their commitments related to financial resources and transfer of technology, considering that 'economic and social development and poverty eradication are the first and overriding priorities of the developing country Parties'.[146] The binary understanding of differentiation finds its fullest expression in the Kyoto Protocol, which imposes emissions reduction targets only on developed countries. From the outset, the Kyoto Protocol's endorsement of such differentiation in favour of developing countries proved deeply contentious and was a major factor that influenced the decision of the United States not to ratify the treaty.[147] Further, there have been deeply divergent interpretations of the scope of the CBDRRC principle and how it ought to be applied in practice.[148]

142 Ministerial Declaration of the Second World Climate Conference, 6–7 November 1990, para. 5.

143 Lavanya Rajamani, 'The Principle of Common but Differentiated Responsibility and the Balance of Commitments under the Climate Regime' (2000) 9 *Review of European Community and International Environmental Law* 120, pg. 121.

144 Ibid.

145 Ibid; Lavanya Rajamani, 'Differentiation in the Emerging Climate Regime' (2013) 14 *Theoretical Inquiries in Law* 151, pg. 152.

146 Article 4(7) of the UNFCCC.

147 The US Senate unanimously passed the Byrd–Hagel Resolution (US Senate Resolution 98, 105th Congress, 1st session, 25 July 1997), which states that the Kyoto Protocol's exemption for developing countries is environmentally flawed and that differentiation could result in serious harm to the United States economy.

148 For a recent analysis of differing applications of CBDRRC suggested by member states, see Lavanya Rajamani, 'The Reach and Limits of the Principle of Common but Differentiated Responsibilities

Starting with the Bali Action Plan of 2007, which launched a process to reach 'an agreed outcome' on long-term cooperative action, there were signs that the binary notion of differentiation would eventually give way to a broader concept that goes beyond the simple distinction between developed and developing countries.[149] Eventually, in the Paris Agreement, 'the era of strict bifurcation has come to an end',[150] and what we have is '[enhanced] symmetry or parallelism between developed and developing countries'. The Paris Agreement requires all parties (not just developed country parties) to prepare, report, and maintain successive 'nationally determined contributions', or NDCs.[151] Countries are required to update their NDCs every five years.[152] Each update needs to represent 'a progression beyond the Party's then current nationally determined contribution'.[153] It also needs to take into account the five-yearly 'global stock take' exercise mandated under Article 14 of the Paris Agreement to assess 'the collective progress towards achieving the purpose of this Agreement and its long-term goals'. Maljean-Dubois describes the NDC process as a reflection of self-differentiation, which 'is the result of a fully bottom-up (and voluntary) process of self-determination of national pledges'.[154] She further argues that the Paris Agreement embodies a more dynamic notion of differentiation, whereby each section takes a different approach to differentiation, 'carefully balancing what will be differentiated and what will be common in the post-2020 period'.[155] For example, the finance provisions are based on a strong version of differentiation. As such, developed countries 'shall provide financial resources to assist developing country Parties'.[156] On the other end of the spectrum are the provisions pertaining to the transparency framework. Maljean-Dubois points out that it is in this part of the Paris Agreement that 'the obligations of developed and developing countries are converging the most' as all 'Parties shall account for their nationally determined contributions'[157] even if the transparency framework takes into account parties' different capacities. These developments mark a significant departure from the differential treatment based on the binary distinction drawn between developed and developing countries.

As the brief discussion here shows, the CBDRRC principle is an important normative pillar of the international climate change regime. Its meaning and application have changed significantly over time, and some may say that such

and Respective Capabilities in the Climate Change Regime' in Navroz K. Dubash (ed), *Handbook of Climate Change and India: Development, Politics and Governance* (Earthscan 2012).

149 Charlotte Streck, Paul Keenlyside, and Moritz von Unger, 'The Paris Agreement: A New Beginning' (2016) 13 *Journal for European Environmental and Planning Law* 3, pg. 7.

150 Ibid., pg. 13.

151 Article 14(2) of the Paris Agreement; see ibid. pg. 11 for discussion on the NDCs.

152 Article 4(9) of the Paris Agreement.

153 Article 4(3) of the Paris Agreement

154 Sandrine Maljean-Dubois, 'The Paris Agreement: A New Step in the Gradual Evolution of Differential Treatment in the Climate Regime?' (2016) 25 *Review of European Community and International Environmental Law* 1, pg. 4.

155 Ibid., pg. 2.

156 Article 4 of the Paris Agreement.

157 Article 4(13) of the Paris Agreement.

changes constitute refinements that will bode well for the development of a comprehensive international framework on climate change. Cities have also taken the CBDRRC principle into account in their norm-setting actions. The preamble of the Seoul Declaration issued at the C40 Large Cities Climate Summit 2009 states that C40 cities share the view that it is necessary to take immediate actions 'based on the principles of co-existence, mutual benefit and common but differentiated responsibilities'.[158] Accordingly, 'cities in developed countries need to assist the efforts of cities in developing countries',[159] and leadership is expected from more-developed and wealthier cities that have the resources, for example, to organize and host summits, conferences, and workshops. C40 is not a signatory to the climate change treaties and therefore does not face any legal obligations to abide by the CBDRRC principle. Its adoption of this principle, and application to its member cities, is a voluntary act that indirectly expands the authority of the UNFCCC regime. It should be noted that there are C40 cities located in states that did not ratify the Kyoto Protocol.[160] Further, C40 member cities Toronto and Vancouver are located in Canada, a state that withdrew from the Kyoto Protocol during its second commitment period.[161] Therefore, it can be argued that prior to the signing of the Paris Agreement, C40's application of the CBDRRC principle was a way for UNFCCC norms to circumvent recalcitrant or reluctant national governments and find articulation within states at the sub-national level.

It should be noted that although the CBDRRC principle has traditionally been applied in the context of the multilateral climate change negotiations, it can have wider application in the transnational climate change regime complex. For example, Scott and Rajamani have argued that the fact that the CBDRRC principle is a 'fundamental part of the conceptual apparatus of the climate change regime also implies ... that state parties are obliged not just to interpret current obligations and fashion new ones in keeping with the CBDRRC principle, but also to take this principle into account in their unilateral actions *vis-à-vis* other parties'.[162] C40's endorsement of the CBDRRC principle similarly indicates the principle's wider relevance, which in the present case constitutes horizontal application of the principle amongst sub-state entities domestically and globally.

6.4.3 *MRV and Transparency*

The architects of the Paris Agreement intend to use data transparency as a driving force to build trust and create incentives for parties to work towards climate

[158] Preamble of the Seoul Declaration, adopted at the Third C40 Large Cities Climate Summit, Seoul, South Korea, 21 May 2009.

[159] Ibid. [160] For example, the United States.

[161] Canada withdrew from the Kyoto Protocol on 15 December 2012; UNFCCC, 'Status of Ratification', online: unfccc.int/kyoto_protocol/status_of_ratification/items/2613.php (accessed on 1 July 2016).

[162] Joanne Scott and Lavanya Rajamani, 'EU Climate Change Unilateralism' (2012) 23 *European Journal of International Law* 469, pg. 477.

mitigation based on domestically determined targets volunteered on a bottom-up basis.[163] Article 13 of the Paris Agreement establishes 'an enhanced transparency framework for action and support'. This information-based mechanism embodies the approach whereby GHG emission reduction targets are determined nationally, whereas MRV is organized at the international level. In order to 'build mutual trust and confidence and to promote effective implementation',[164] the transparency framework is intended to provide informational clarity and permit tracking of individual states' progress towards achieving their NDCs. Each party is required to regularly provide a national inventory report of its GHG emissions and 'information necessary to track progress made in implementing and achieving its nationally determined contribution under Article 4'.[165] Article 13 is therefore described as relying on a form of 'naming and shaming' to nudge states into taking action not only in connection with mitigation and adaptation, but also in relation to assistance.[166] The information submitted by states will be subject to a technical expert review, which will identify areas of improvement and capacity-building needs amongst other things.[167] From the outset, India and China were strongly opposed to an international MRV mechanism because of its perceived intrusiveness.[168] At the crux of their reservations was the extent of differentiation in relation to transparency.[169] As a result, the final wording of Article 13 reflects a degree of compromise and gives assurance that the transparency framework will be 'implemented in a facilitative, non-intrusive, non-punitive manner, respectful of national sovereignty, and avoid placing undue burden on Parties'.[170]

The extensive reliance on information-driven techniques to implement the Paris Agreement is the first of its kind in international environmental law. To ensure that this governance experiment gets on the right track to success requires the energy and contribution of various actors in the transnational climate change regime complex. In this regard, urban climate law complements and supports the information-driven approach embodied in the Paris Agreement. In Chapter 5, I introduced the Compact of Mayors. Briefly, it is a voluntary scheme that certifies cities that have fulfilled the compliance requirements, which primarily involve accounting and reporting of GHG emission inventories as well as mitigation and adaptation actions.

[163] For an overview of differing opinions on whether the Paris Agreement represents a step forward in addressing climate change given that 'country specific targets volunteered on a bottom-up basis are less likely to be sufficiently ambitious, in the aggregate, to meet global goals', see Streck, Keenlyside, and von Unger, pg. 28.

[164] Article 13(1) of the Paris Agreement. [165] Article 13(7) of the Paris Agreement.

[166] Jorge E. Viñuales, *The Paris Climate Agreement: An Initial Examination (Part III of III)* (Blog of the *European Journal of International Law*, 8 February 2016), online: www.ejiltalk.org/the-paris-climate-agreement-an-initial-examination-part-iii-of-iii/ (accessed on 1 July 2016).

[167] Article 13(11) of the Paris Agreement.

[168] Alex Barker and Pilita Clark, 'India Slows Progress on Ambitious Climate Change Accord' *Financial Times* (16 November 2015); Pilita Clark, 'COP21: China Accused of Blocking Progress at Paris Climate Talks' *Financial Times* (8 December 2015).

[169] Viñuales. [170] Article 13(3) of the Paris Agreement.

It is noteworthy that, at the launch of the Compact of Mayors at the high-profile Climate Summit in New York City on 23 September 2014, it was repeatedly emphasized that one of the Compact's key aims is to '[d]emonstrate the commitment of city governments to ... more ambitious, transparent, and credible national climate targets by *voluntarily agreeing to meet standards similar to those followed by national governments*' (emphasis added) in the run-up to the UNFCCC 21st COP in Paris and beyond.[171] Specifically, the Compact seeks to have city-level initiatives complement and support the international climate negotiations by adopting the pledge-and-review approach that underpins the Paris Agreement and developing a set of MRV standards and practices that are similar to those at the international level. For example, the 'City Climate Commitments'[172] that cities are required to pledge upon joining the Compact are similar to the NDCs that states are required to report pursuant to the Paris Agreement. In both cases – the City Climate Commitments and the NDCs – it is expected that civil society, journalists, and other interested actors who are able to scrutinize the publicly available data will contribute to holding cities and states accountable. The hope is that the public scrutiny will create incentives for city and national governments to undertake more ambitious actions to secure reputational benefits or avoid being 'named and shamed' for deviating from the standard of appropriate behaviour. At the time of writing, the modalities, procedures, and rules that will underpin the Paris Agreement's transparency framework have yet to be determined. It is therefore not possible to undertake a more detailed analysis of MRV at the international level and the sub-national level except to note that there are already clear signs that cities intend to develop a global transparency framework that will mirror MRV at the international level as closely as possible. In this third example of coupling, we see how urban climate law seeks to complement and strengthen the international climate regime.

6.5 CONCLUSION

This chapter sought to advance the argument that global cities are beginning to perform a role in creating law. Based on a set of norms that posit that cities ought to and can play a meaningful role in global efforts to address climate change, global cities have cooperated with various actors, including development banks, NGOs, and environmental consultancies to develop practices and voluntary standards. I coin the term 'urban climate law' to refer to the norms, practices, and voluntary standards developed by global cities and implemented through their transnational networks.

[171] Cities Mayors Compact Action Statement (Issued at the Climate Summit 2014), 'Goals, Objectives and Commitments', online: www.un.org/climatechange/summit/wp-content/uploads/sites/2/2014/09/CITIES-Mayors-compact.pdf (accessed on 1 July 2016).

[172] Compact of Mayors, online: www.compactofmayors.org/resources/ (accessed on 1 July 2016).

Urban climate law relies on two key pathways to steer the behaviour of cities and their authorities towards climate mitigation and adaptation as well as investment in low-carbon development options for the future. The promotion of reflexivity and norm diffusion often do not proceed in a linear fashion but constitute reiterative, dynamic interactional processes to spread the adoption of norms, practices, and voluntary standards. Relying on insights from the theoretical literature and empirical findings, the discussion in this chapter showed how cities and their governments engage in learning and imitation, and emphasized the crucial role of norm entrepreneurs in facilitating these processes.

Finally, the last part of this chapter advanced the argument that the interaction between urban climate law and the UNFCCC regime has been strategic and mostly mutually beneficial. Coupling not only strengthens the UNFCCC regime; endorsement and affirmation of the principles and practices of the UNFCCC also lends legitimacy to urban climate law. The analysis on coupling also makes a contribution to the enduring debate about the role of soft law in global governance by demonstrating that urban climate law, as a form of soft law, plays an important complementary role to that of the relatively hard UNFCCC regime through elaborating on norms, facilitating experimentation, and generating knowledge about novel institutional practices. As consistently argued in this chapter, soft law and hard law not only support and complement one another, but the interaction between them can also result in a synergistic expansion of governance that adds to the overall coherence of the transnational climate change regime complex. It can be argued that the international community sacrificed ambition for the sake of universal participation when it agreed to put the NDCs at the centre of the Paris Agreement's mitigation framework. As a voluntary, bottom-up process, there is the risk that the aggregate effect of the NDCs will not be sufficient to meet the two-degree Celsius target. There is already evidence of this ambition gap, as the Paris Decision 'noted with concern'.[173] Therefore, it is all the more important that other actors in the transnational climate change regime complex help maintain the momentum and ambition that was seen in the run-up to COP-21 in Paris and also support the development of the novel mechanisms that the Paris Agreement has introduced. If the Paris Agreement is a milestone that marks the beginning of a new era for the international climate regime, it also marks the beginning of a new chapter for the transnational climate change regime complex. Within the regime complex, global cities have a unique contribution to make.

[173] Para. 17, Decision 1/CP. 21: Adoption of the Paris Agreement, UNFCCC, Report of the Conference of the Parties on its twenty-first session, held in Paris from 30 November to 13 December 2015.

7

A Normative Assessment of Urban Climate Law

7.1 INTRODUCTION

The preceding chapters examined in detail what five global cities are doing to address climate change and how global cities have come together to form a transnational network to scale up their climate actions and facilitate the diffusion of norms, practices, and voluntary standards. Chapter 6 examined the mechanisms of urban climate law and its interactions with the UNFCCC normative framework. This chapter endeavours to take a step back from the intricate details and consider some 'big picture' questions. The first question I would like to explore in this chapter is how the transnational cooperative efforts among global cities and urban climate law contribute towards the performance of the transnational climate change regime complex.

Throughout this book, Abbott's conception of the transnational climate change regime complex has been used to frame the discussion of how multiple governance actors and institutions currently govern climate change. As mentioned in Chapter 2, Abbott's definition is a refinement of Keohane and Victor's conceptualization of the climate change regime complex in their article published in 2011 as 'a loosely coupled system of institutions ... [which] are linked in complementary ways'.[1] In this framing, urban climate law constitutes a regime or institution within the regime complex. Keohane and Victor offer six evaluative criteria to assess regime complexes normatively and also identify some ways in which the functioning of the climate change regime complex can be improved. These six criteria are: (1) coherence amongst regimes in the sense of being compatible and mutually reinforcing, (2) accountability to relevant audiences, (3) determinacy of rules in order to enhance compliance and reduce uncertainty, (4) sustainability in the sense of being durable, (5) epistemic quality in rules, and (6) fairness in the sense that '[institutions] should provide benefits widely'.[2] The choice of these criteria is not random; factors such as

[1] Robert O. Keohane and David G. Victor, 'The Regime Complex for Climate Change' (2011) 9 *Perspectives on Politics* 7, pg. 9.
[2] Keohane and Victor, pgs. 16–17.

coherence and the epistemic quality of rules are widely used as benchmarks to evaluate the integrity and soundness of normative systems. In Section 7.2 of this chapter, these six criteria will be used to evaluate the extent to which urban climate law contributes towards enhancing the overall performance of the transnational climate change regime complex. The key conclusion is that the involvement of global cities and their networks strengthens the transnational climate change regime complex.

It should be clarified that the aim of Section 7.2 is not to assess the overall performance of the transnational climate change regime complex but rather to evaluate the normative contribution that global cities make – through their creation and implementation of urban climate law – to the transnational governance of climate change. As Keohane and Victor have pointed out, '[w]hether the proliferation of different forums working on the climate issue – such as the G20, the MEF, various bilateral technology and investment partnerships, and private sector and NGO initiatives – is an asset or liability depends on their content and how these efforts are coupled'.[3] In carrying out the assessment in Section 7.2, I will also elaborate upon the six criteria to provide a fuller basis for future regime complex analysis.

Having concluded in Section 7.2 that global cities and their networks have a valuable normative role to play in the transnational climate change regime complex, Section 7.3 takes a step further in the macro-level analysis and considers what the emergence of cities as transnational lawmaking actors means for the study of international law and international relations more broadly. While the growing prominence of cities at the international level has generated lively debate amongst sociologists, political scientists, and governance scholars, international law scholars have largely neglected this development.[4] It will seem that what is unfolding in practice – that cities perceive themselves as having a role to play in global governance, that the transnational activities of cities generate normativity, and that regimes such as the UNFCCC recognize cities as stakeholders in the international climate change negotiations – is not relevant to international law. This book has sought to show that the opposite is true. For one, the transnational governance activities of

[3] Ibid.

[4] There are a few notable exceptions. See, for example, Yishai Blank, 'Localism in the New Global Legal Order' (2006) 47 *Harvard International Law Journal* 263; G. E. Frug and David Barron, 'International Local Government Law' (2006) 38 *Urban Lawyer* 1; Ileana Porras, 'The City and International Law: In Pursuit of Sustainable Development' (2008) 36 *Fordham Urban Law Journal* 537; Janne E. Nijman, 'Renaissance of the City as Global Actor: The Role of Foreign Policy and International Law Practices in the Construction of Cities as Global Actors' in Gunther Hellmann, Andreas Fahrmeir, and Milos Vec (eds), *The Transformation of Foreign Policy: Drawing and Managing Boundaries from Antiquity to the Present* (Oxford University Press 2016); Janne E. Nijman, 'The Future of the City and the International Law of the Future' in Sam Muller et al. (eds), *The Law of the Future and The Future of Law* (Torkel Opsahl Academic EPublisher (Oslo, Norway) 2011); Helmut Philipp Aust, 'Shining Cities on the Hill? The Global City, Climate Change, and International Law' (2015) 26 *European Journal of International Law* 255.

cities ought to be of interest to international law scholars because these activities steer behaviour in a law-like manner. Further, as the analysis in Section 7.2 of this chapter will show, the norms, practices, and voluntary standards that global cities have developed and are implementing raise questions of accountability, transparency, and fairness. These are questions that are central to the study of international law in today's world. There is therefore a case to be made for international law scholars paying attention to the rise of global cities as relevant actors in the climate change context and in international affairs more broadly. Section 7.3 will also offer some thoughts on how the responses of international law and practice to the emergence of global cities as transnational lawmaking actors may contribute towards the ongoing multidisciplinary conversation about global cities. Section 7.4 offers some concluding remarks.

7.2 EVALUATING URBAN CLIMATE LAW IN THE TRANSNATIONAL REGIME COMPLEX CONTEXT

A key question that arises from the discussion in earlier chapters is whether the normative activities of global cities make an overall positive contribution towards the collective efforts of states and non-state actors to govern climate change more effectively and therefore bring us closer to achieving the objectives of the Paris Agreement. In their pioneering article on the climate change regime complex, Keohane and Victor propose six criteria as a framework for analyzing the performance of the climate change regime complex as a whole. This section posits that it would be meaningful to use the criteria to evaluate the performance of the individual regimes or institutions within the regime complex. Doing so presents one way of answering the question of whether urban climate law, as one of the regimes within the climate change regime complex, contributes positively to transnational climate change governance and what improvements may be made.

In the discussion to follow, each criterion will be elaborated upon before it is used to evaluate salient aspects of the creation and implementation of urban climate law. This section will conclude with a summary overview before the discussion in Section 7.3 about the broader implications that the rise of global cities as lawmaking actors poses for the study of international law.

7.2.1 *Coherence*

In Keohane and Victor's view, the coherence of a regime complex depends on the extent to which the various specific regimes are 'compatible and mutually reinforcing'.[5] Coherence is perceived to be a good thing in a utilitarian sense: '[w]here compatibilities exist they encourage linkages that make it easier to channel

[5] Keohane and Victor, pg. 16.

resources from one element of the regime complex to another'.[6] I will seek to furnish Keohane and Victor's definition with more detail before proceeding to evaluate the extent to which urban climate law contributes towards building coherence in the regime complex.

The concept of coherence has received significant attention in the literature on the functioning of the EU, which seeks to live up to its commitment to ensure the coherence of its policies. Interest in coherence predated the creation of the EU's single institutional framework, given that, without the unifying framework, the political functioning of the EU rested upon the legal obligation of coherence.[7] Today, Article 13 of the Treaty on European Union constitutes the legal basis for coherence in EU foreign policy, and coherence has emerged as a principle understood to impose a procedural obligation on EU foreign policy actors to coordinate their policies.[8] From the perspectives of the European Commission and the Council, a coherent EU foreign policy is a necessary precondition for effectiveness.[9] Several scholars have pointed out the distinction between consistency and coherence. In Hillion's view, for example, coherence goes beyond the assurance that different policies do not contradict each other, to seek synergy and added value in the different components of EU policies.[10] Hoffmeister also argues that the notion of coherence relates more to the pursuit of positive synergies than simply avoiding policy contradictions.[11] Finally, it has been argued that consistency is a binary definition – that is, policies in question are consistent or not. In contrast, we can conceive of a spectrum of coherence whereby a set of policies can be more or less coherent.[12] In brief, coherence can take the form of a legal obligation of a procedural nature to avoid inconsistencies in a set of policies, but it is widely argued that coherence is more than seeking consistency. It relates to seeking positive synergies to maximize policy benefits.

6 Ibid.

7 Deidre Curtin, 'The Constitutional Structure of the Union: A Europe of Bits and Pieces' (1993) 30 *Common Market Law Review* 17, pg. 27.

8 Clara Portela and Kolja Raube, *(in-)Coherence in EU Foreign Policy: Exploring Sources and Remedies* (Paper presented at the European Studies Association Biannual Convention, Los Angeles, April 2009), pg. 4. Article 13 of the Treaty on European Union states, 'The Union shall have an institutional framework which shall aim to promote its values, advance its objectives ... and ensure the consistency, effectiveness and continuity of its policies and actions.'

9 See, for example, European Commission, *Europe in the World – Some Practical Proposals for Greater Coherence, Effectiveness and Visibility* [Communication from the Commission to the European Council of June 2006 COM (2006) 278 final, 2006].

10 Christophe Hillion, 'Tous pour un, Un pour tous! Coherence in the External Relations of the European Union' in M. Cremona (ed), *Developments in EU External Relations Law, Collected Courses of the Academy of European Law* (Oxford University Press 2008), pg. 17.

11 Frank Hoffmeister, 'Inter-Pillar Coherence in the European Union's Civilian Crisis Management' in Steven Blockmans (ed), *The European Union and Crisis Management: Policy and Legal Aspects* (T.M.C. Asser Press 2008), pg. 161.

12 Antonio Missiroli, *Coherence for European Security Policy: Debates, Cases, Assessments* (Occasional Papers 27, Institute for Security Studies, Western European Union, 2001), pg. 4; online: www.iss.europa.eu/uploads/media/occ027.pdf (accessed on 29 July 2016).

To recall its definition, a regime complex is a loosely coupled system of institutions. It is characterized by the absence of hierarchy amongst the institutions, and there is little, if any, central coordination amongst institutions. Thus, unlike an institutional complex like the EU, regime complexes do not have formal institutional arrangements and processes in place to create coherence. Instead, coherence in a regime complex has to be orchestrated through less-formal efforts by the actors involved. The literature on regime interplay, with its focus on the goals and methods to achieve integration in a regime complex, offers some relevant insights. For example, coherence can be enhanced through the adherence to overarching goals such as avoiding conflict, enhancing synergy, achieving efficiency, and promoting equity.[13] Coordination through markets and networks are some of the methods that can be employed to create integration and coherence. Further, Morin and Orsini remind us that '[t]he creation and development of regime complexes is anything but a natural process; they are actively constructed by agents'. They contend that the more coherent states and other actors are in their policies towards the specific regimes in a regime complex, the more likely they are to promote 'density' in the regime complex.[14] 'Density' refers to the number of connections between the regimes, and the assumption is that the greater the number of connections between regimes, the more pathways and opportunities exist for the fostering of coherence amongst the norms and practices of the individual regimes.

The literature on regime interaction (denoting that one regime may influence others[15]) is also relevant to the study of coherence. Oran Young was one of the first scholars to identify different types of regime interactions, including 'clustered institutions' (where several indirectly related institutions are bundled together, as has happened in the 'package deals' that formed the World Trade Organization and the United Nations Convention on the Law of the Sea) and 'overlapping institutions' (where distinct institutional arrangements affect each other in mostly unintentional ways – for example, the trade and environmental agreements).[16] Stokke refined Oran Young's typology by introducing the notions of 'normative interaction' (where one regime may confirm or contradict the norms of another) and 'ideational interaction' (where one regime may learn from another).[17] The consequences of regime interaction can include conflicts or synergies, and it is the latter that I am interested in exploring for present purposes.

[13] Sebastian Oberthür, 'Interplay Management: Enhancing Environmental Policy Integration among International Institutions' (2009) 9 *International Environmental Agreements: Politics, Law and Economics* 371.

[14] Jean Frédéric Morin and Amandine Orsini, 'Policy Coherency and Regime Complexes: The Case of Genetic Resources' (2014) 40 *Review of International Studies* 303, pg. 308.

[15] H. van Asselt, *The Fragmentation of Global Climate Governance: Consequences and Management of Regime Interactions* (Edward Elgar 2014), pg. 46.

[16] Oran Young, 'Institutional Linkages in International Society: Polar Perspectives' (1996) 2 *Global Governance* 1, pgs. 2–6.

[17] Olav Schram Stokke, *The Interplay of International Regimes: Putting Effectiveness Theory to Work* (Fridtjof Nansen Institute, FNI Report 14/2001, 2001), pg. 10.

For Kristen Rosendal, there is synergy when the aggregate effects of two institutions are larger than the sum of effects produced on their own.[18] As van Asselt puts it, 'the term has a positive connotation, associated with enhancing the effectiveness of one or both interacting regimes'.[19] Van Asselt provides a list of indicators to assess whether regime interaction could lead to synergies, which can usefully be adapted to determine the existence of synergies in a regime complex.[20] The first indicator is whether there are shared principles, the assumption being that when regimes apply the same principles, such as the CBDRRC principle, it enhances the regime complex's coherence. The second indicator is the existence of common economic incentives to promote the same type of activities. Van Asselt gives the example of the Global Environmental Facility serving as the financial mechanism for a number of multilateral agreements, thereby providing common economic incentives to address global environmental problems including biodiversity loss and climate change. The third indicator is the existence of streamlined monitoring and reporting obligations. Van Asselt points out that synergies are created when the monitoring and reporting obligations under different regimes are streamlined, thereby reducing data collection costs. The fourth indicator is the existence of shared supporting measures such as capacity building and technology transfer mechanisms amongst the regimes. The fifth indicator is the extent of learning among the regimes.

Applying the first indicator of shared norms and principles, it can be argued that urban climate law contributes to consistency and compatibility because the norms and practices of cities are consistent with the goals of other actors in the regime complex – climate mitigation and adaptation – and the general adherence to the notions of environmental liberalism. Furthermore, cities acting through C40 support the CBDRRC principle. Another example is the Compact of Mayors initiative, which is intended to complement the transparency framework engendered by the Paris Agreement. Therefore, there is strong adherence by global cities to a common principle of transparency that forms a key pillar of the post-2020 UNFCCC regime.

With regard to the second indicator of whether common economic incentives to promote the same type of activities exist, it can be argued that climate finance in the UNFCCC regime and other institutions such as the World Bank and development aid agencies provides common economic incentives to pursue low-carbon development, climate mitigation, and adaptation. The common economic incentives provided by climate finance are being extended to the realm of sustainable urban development, and global cities are being encouraged to engage in transnational climate financing arrangements in their efforts to address climate change. As discussed in earlier chapters, the various 'orchestration' efforts by the World

18 Kristen G. Rosendal, 'Impacts of Overlapping International Regimes: The Case of Biodiversity' (2001) 7 *Global Governance* 95, pg. 97.

19 Van Asselt, pg. 55.

20 Ibid., pgs. 56–58. The remainder of this paragraph draws from van Asselt unless otherwise indicated.

Bank and other actors include training programmes to build the capacity of city governments to gain access to sources of climate finance and 'matchmaker workshops' to connect potential investors and city governments that are seeking funding for climate projects.

The streamlining of reporting and monitoring obligations creates further synergy between urban climate law, climate initiatives by private actors, and the UNFCCC regime. The UNFCCC has a global platform known as the *Non-State Actor Zone for Climate Action* (NAZCA), which tracks climate action commitments by companies, cities, regions, and investors.[21] The information that cities report via the Compact of Mayors, CDP, and carbon*n* Climate Registry is automatically included in NAZCA.[22] This enhances transparency. At the same time, a readily accessible platform for climate change governance actors to learn about potential collaboration opportunities has the potential to enhance synergistic cooperation and complementary action amongst global cities and other actors in the regime complex.

The fifth indicator points to the extent of learning among the regimes. Cities contribute to the learning processes within the regime complex when they cooperate with other institutions in developing their norms, practices, and voluntary standards. This has been discussed in detail in Chapter 6 and will not be reiterated here.

To summarize, the de minimis requirements of coherence are the consistency and compatibility of the practices and outputs of the various regimes. Synergies among regimes in a regime complex create coherence, whereby synergy is said to exist when the aggregate effects of two regimes are larger than the sum of effects produced on their own.[23] A regime complex can be said to be more coherent when there is more synergy amongst the regimes within it, and less coherent when there is less synergy. Finally, we can use a number of indicators, as suggested by van Asselt, to gauge the level of synergy, thereby also gauging the amount of coherence in a regime complex. It can be argued that urban climate law is largely consistent and compatible with the norms and practices of other institutions in the regime complex. Particularly in relation to shared norms, streamlined monitoring and reporting obligations, and the extent to which global cities contribute towards learning within the regime complex, global cities can be said to be enhancing synergies and therefore contributing positively to the amount of coherence in the transnational climate change regime complex.

7.2.2 *Determinacy*

Keohane and Victor draw their determinacy criterion from Thomas Franck's work on legitimacy. In Franck's view, there are four characteristics of a rule that determine its degree of legitimacy: determinacy, symbolic validation, coherence, and

[21] NAZCA, online: http://climateaction.unfccc.int (accessed on 1 August 2016).

[22] NAZCA, 'Data Partners', online: http://climateaction.unfccc.int/about (accessed on 1 August 2016).

[23] Rosendal, pg. 97.

adherence.[24] To the extent that a rule exhibits these four characteristics, they exert an inherent power over states to comply with the rule in question.[25] Of the four characteristics, Franck argues that textual determinacy is '[p]erhaps the most self-evident of all characteristics making for legitimacy'.[26] Textual determinacy refers to 'the ability of the text to convey a clear message' so that those addressed know exactly what is required of them, which is an essential prerequisite for compliance.[27] 'Readily ascertainable normative content' is therefore another aspect of determinacy.[28]

Franck argues that the indeterminacy of rules can be costly in the sense that indeterminacy makes it easier for actors to justify non-compliance.[29] Conversely, a more determinate rule leaves less room for interpretation and therefore is less amenable to evasive strategies that are employed to justify non-compliance. Franck also points out that the degree of the determinacy of a rule directly affects the degree of its perceived legitimacy.[30] This point is illustrated as follows: 'A rule that prohibits the doing of "bad things" lacks legitimacy because it fails to communicate what is expected, except within a very small constituency in which "bad" has achieved a high degree of culturally induced specificity. To be legitimate, a rule must communicate what conduct is permitted and what conduct is out of bounds.'

While Franck is concerned with the textual determinacy of rules found in treaties, resolutions of international organizations, and so forth, nothing in his analysis precludes extending the discussion of rule determinacy to the realm of voluntary standards.[31] In a community organized around rules, whether it is a community of states, private actors, or cities, compliance is secured, at least in part, by the perception of a rule as legitimate by those to whom it is addressed. Whether the rule is found in customary international law or is promulgated by a voluntary certification scheme does not really make a difference; the question of whether those to whom the rule is addressed will comply turns on whether they deem it legitimate. Legitimacy, in turn, is partly determined by the textual determinacy of the rule in question.

As discussed in previous chapters, the GHG Protocol for Cities and the Compact for Mayors both provide users with detailed user guidelines that are intended to confer a high degree of specificity on the accounting standards and criteria that global cities are required to apply. The Compact of Mayors also publishes user

[24] Thomas M. Franck, 'Legitimacy in the International System' (1988) 82(4) *American Journal of International Law* 705, pg. 712. Franck defines 'legitimacy' to mean 'that quality of a rule *which derives from a perception on the part of those to whom it is addressed that it has come into being in accordance with right process*'; pg. 706.

[25] Ibid., pg. 712. [26] Ibid., pg. 713. [27] Ibid.

[28] Thomas M. Franck, *The Power of Legitimacy among Nations* (Oxford University Press 1990), pg. 52.

[29] Franck, 'Legitimacy in the International System', pg. 714. [30] Ibid., pg. 716.

[31] Naiki makes this point in assessing the sustainable bioenergy regime complex; Yoshiko Naiki, 'Trade and Bioenergy: Explaining and Assessing the Regime Complex for Sustainable Bioenergy' (2016) 27 *European Journal of International Law* 129, pg. 151.

guides that are designed to help cities gain a firm grasp of the scope of their reporting obligations, thereby making the normative content of the voluntary standard clearly ascertainable to cities that seek certification.[32] The GHG Protocol for Cities is contained in a document of over 170 pages consisting of detailed information about goal setting, determining the boundaries or parameters of the city's GHG inventory, the accounting methodologies for various sectors (such as waste and transportation), and methods of verification to assess the completeness and accuracy of reported data.[33] GHG emission standards enjoy a high degree of determinacy because they lend themselves to objective measurement. Thus, it can be argued that the voluntary standards and practices that global cities have developed and implemented contribute positively towards determinacy in the transnational regime complex for climate change because of their high level of textual specificity and clarity in normative content. However, it should be noted that as far as voluntary standards are concerned, epistemic quality and scientific knowledge may be more important considerations. Naiki argues, 'even ... [voluntary] standards that appear to be determinate may not be considered legitimate and credible if they are not based on epistemic quality and scientific knowledge'.[34] It is to the dimension of epistemic quality that the discussion turns next.

7.2.3 *Epistemic Quality*

Keohane and Victor do not say much about epistemic quality except that one aspect of epistemic quality is 'the consistency between ... rules and scientific knowledge' and that the epistemic quality of a regime complex bears importance for its legitimacy and effectiveness.[35] 'Epistemic quality' is generally defined as 'reliable information needed for grappling with normative disagreement and uncertainty'[36] – such information being provided by experts. As modern societies grapple with complex risks pertaining to food safety, chemicals, the stability of financial systems, and other domains, they are increasingly dependent on expert knowledge because '[o]ne cannot regulate what one does not understand'.[37] As the risks that require regulation are usually complex and rapidly changing, regulators themselves need to be highly skilled or rely on experts.[38] Peter Haas has explored the role of experts in global

[32] Compact of Mayors, 'Resources', online: www.compactofmayors.org/resources/tools-for-cities/ (accessed on 1 August 2016).

[33] *Greenhouse Gas Protocol, Global Protocol for Community-Scale Greenhouse Gas Emission Inventories: An Accounting and Reporting Standard for Cities* (World Resources Institute, C40, and ICLEI 2014).

[34] Naiki, pg. 152. [35] Keohane and Victor, pg. 17.

[36] Allen Buchanan and Robert O. Keohane, 'The Legitimacy of Global Governance Institutions' (2006) 20 *Ethics & International Affairs* 405, pg. 426.

[37] Martin Shapiro, 'Deliberative, Independent Technocracy v. Democratic Politics: Will the Globe Echo the E.U.?' (2005) 68 *Law and Contemporary Problems* 341, pg. 343.

[38] In the EU, where the European Commission routinely consults around a thousand expert groups and there are currently more than forty increasingly powerful EU agencies, fundamental questions have

environmental governance extensively in his work, coining the term 'epistemic communities' to describe the networks of 'knowledge-based experts' that play a role in 'articulating the cause-and-effect relationships of complex problems, helping states identify their interests, framing the issues for collective debate, proposing specific policies, and identifying salient points for negotiation'.[39]

Epistemic quality bears particular salience for voluntary standards because they derive part of their legitimacy from claims of expert knowledge. Within the literature on voluntary standards, scholars with an institutionalist perspective have focused much of their efforts on identifying how standards and the organizations behind them achieve legitimacy.[40] One of the central tenets of this body of literature is that standard-setting entities rely on expertise to build authority and legitimacy such that standards are made out to be 'expert knowledge in the form of rules'.[41] Legitimacy has been defined as 'the justification of actions to those whom they affect according to reasons they can accept'.[42] There are two aspects of legitimacy: input (process) and output (performance and effectiveness) legitimacy. Bodansky explains that the former 'derives from the process by which decisions are made, including factors such as transparency, participation, and representation', whereas the latter concerns effectiveness and 'the results of governance'.[43] Kerwer posits that a necessary (but not always sufficient) precondition for the effectiveness of a standard is that the target audience believes that the expertise on which the standard is based is convincing.[44]

While expert knowledge can stem from many disciplines, 'science has an institutionalized monopoly of knowledge' in contemporary societies.[45] Science is

been raised about the legitimate role of knowledge and expertise in decision-making. Critics claim that growing expert power and the increase in depoliticized (and therefore politically unaccountable) expert bodies erode democratic government; *Expertise and Democracy* (Centre for European Studies, University of Oslo, ARENA Report 1/14, edited by Cathrine Holst, 2014), pgs. 2 and 3.

39 Peter M. Haas, 'Introduction: Epistemic Communities and International Policy Coordination' (1992) 46 *International Organization* 1, pg. 2.

40 Stefano Ponte and Emmanuelle Cheyns, 'Voluntary Standards, Expert Knowledge and the Governance of Sustainability Networks' (2013) 13 *Global Networks* 459, pg. 463.

41 Bengt Jacobsson, 'Standardization and Expert Knowledge' in Nils Brunsson and Bengt Jacobsson (eds), *A World of Standards* (Oxford University Press 2002), pg. 41.

42 Melissa S. Williams, 'Citizenship as Agency within Communities of Shared Fate' in Steven Bernstein and William D. Coleman (eds), *Unsettled Legitimacy: Political Community, Power, and Authority in a Global Era* (UBC Press 2009), pg. 43.

43 Daniel Bodansky, 'Legitimacy in International Law and International Relations' in Jeffrey L. Dunoff and Mark A. Pollack (eds), *Interdisciplinary Perspectives on International Law and International Relations* (Cambridge University Press 2012), pg. 330.

44 Dieter Kerwer, 'Rules That Many Use: Standards and Global Regulation' (2005) 18 *Governance* 611, pg. 618. Perez has argued that with declining trust in experts and their professed expertise, the power of 'expert knowledge to provide privileged accounts of the common good and, hence, to serve as a source and arbiter of legitimacy' has also declined. As a result, 'the legitimacy of transnational regimes is judged, increasingly, by the nature of the process that led to the regimes' creation, and by the public accountability of those who implement them'; Oran Perez, 'Normative Creativity and Global Legal Pluralism: Reflections on the Democratic Critique of Transnational Law' (2003) 10 *Indiana Journal of Global Legal Studies* 25, pgs. 28–29.

45 *Expertise and Democracy*, pg. 20.

often portrayed to be objective and capable of eliminating uncertainty, and therefore is used to lend legitimation to policy decisions.[46] However, as the literature on science and technology has shown, science is not produced in a social vacuum and does not discover uncontroverted 'truths' that are then translated into policies or technologies.[47] In fact, facts and values are intertwined, and uncertainty is a constitutive feature of knowledge. Science, however sound, cannot eliminate uncertainty, and the question of how uncertainties are to be managed is a normative one. In spite of these reservations about the objectivity of science, it remains an important aspect of epistemic quality, especially in the case of climate change. There is a widespread perception that science is the final arbiter in the climate change debate, and the IPCC goes to great lengths to establish that it provides assessments based on the best available science to guide governments on international climate policy.[48]

In developing their norms, practices, and voluntary standards, cities have relied on the findings of the IPCC. C40 endorses the IPCC's findings that larger cities consume two-thirds of the world's energy and are responsible for more than three-quarters of global GHG emissions.[49] In addition, cities have sought to ensure that their requirements and guidance for calculating and reporting citywide GHG emissions are consistent with the IPCC guidelines.[50] Therefore, it can be argued that, to the extent that epistemic quality refers to the consistency between urban climate law and IPCC science, urban climate law scores well for this criterion. However, it should also be noted that urban climate law does not only value scientific knowledge as a source of epistemic quality. As illustrated in Chapters 4 and 5, a key characteristic of how cities and their networks function is the formation of extensive partnerships with other actors to tap into their expertise and resources. In developing standards and practices, cities have valued local knowledge and global knowledge. They have therefore sought to tap into the expertise of urban planners, architects, financiers, entrepreneurs, environmental activists, engineers, and development consultants in developing climate solutions.[51]

46 Reiner Grundmann, 'The Role of Expertise in Governance Processes' (2009) 11 *Forest Policy and Economics* 398, pgs. 401–402.

47 See, for example, Sheila Jasanoff and Brian Wynne, 'Science and Decisionmaking' in Steve Rayner and Elizabeth L. Malone (eds), *Human Choice and Climate Change: An International Assessment (Volume 1: The Societal Framework)* (Battelle Press 1998); Sheila Jasanoff (ed) *States of Knowledge: The Co-Production of Science and the Social Order* (Routledge 2004).

48 For discussion about the IPCC, see, for example, Navraj Singh Ghaleigh, 'Science and Climate Change Law – The Role of the IPPC in International Decision-Making' in Cinnamon Carlarne, Kevin Gray, and Richard Tarasofsky (eds), *The Oxford Handbook of International Climate Change Law* (Oxford University Press 2016).

49 See discussion in Section 3.1 of Chapter 3.

50 The GHG Protocol for Cities states that it 'sets out requirements and provides guidance for calculating and reporting city-wide GHG emissions, consistent with the 2006 *IPCC (Intergovernmental Panel on Climate Change) Guidelines for National Greenhouse Gas Inventories*'; pg. 20.

51 See discussion in Chapter 5.

7.2.4 *Accountability*

Keohane and Victor draw on the concept of accountability put forth by Grant and Keohane that 'some actors have the right to hold other actors to a set of standards, to judge whether they have fulfilled their responsibilities in light of these standards, and to impose sanctions if they determine that these responsibilities have not been met'.[52] In other words, the actors being held accountable 'have to answer for [their] action or inaction' concerning 'accepted standards of behavior and ... they will be sanctioned for failures to do so'.[53] Transparency, which has been defined as 'disclosure of information intended to evaluate and/or steer behavior', has been championed as a means of enhancing the accountability of international environmental policy outcomes.[54] As urban climate law involves governance by disclosure, I will focus on the nexus between transparency and accountability in this section.

In considering the consequences of transparency for the accountability and legitimacy of voluntary certification programmes, Auld and Gulbrandsen draw the distinction between procedural and outcome transparency. By 'procedural transparency', they mean 'the openness of governance processes, such as decision-making or adjudication'.[55] It has been argued that procedural transparency is often used to improve the legitimacy of global governance arrangements, which are perceived to be less legitimate because they are not supported by norms of sovereignty and state consent.[56] 'Outcome transparency concerns openness about regulated or unregulated behaviors' and is considered important for identifying and managing environmental problems.[57] Examples include domestic environmental laws requiring industrial facilities to disclose their pollutant discharges to the public and C40 requiring its member cities to report their climate actions on a public register like the carbon*n* Climate Registry.

When it comes to voluntary certification schemes, the target actors for procedural transparency and outcome transparency are the decision-makers and the regulated actors respectively. We will briefly consider each in turn. First, as to procedural transparency, it is argued that the disclosure of information about decision-making

[52] Ruth W. Grant and Robert O. Keohane, 'Accountability and Abuses of Power in World Politics' (2005) 99 *American Political Science Review* 29, pg. 29.

[53] Ibid., at pg. 30.

[54] Aarti Gupta and Michael Mason, 'Transparency and International Environmental Politics' in Michele Betsill, Kathryn Hochstetler, and Dimitris Stevis (eds), *Advances in International Environmental Politics* (Palgrave Macmillan 2014), pgs. 356–357. For discussion, also see Jutta Brunnee and Ellen Hey, 'Transparency and International Environmental Institutions' in Andrea Bianchi and Anne Peters (eds), *Transparency in International Law* (Cambridge University Press 2013).

[55] Graeme Auld and Lars H. Gulbrandsen, 'Transparency in Nonstate Certification: Consequences for Accountability and Legitimacy' (2010) 10 *Global Environmental Politics* 97, pg. 99.

[56] Aarti Gupta argues that this is the central assumption of procedural transparency; Aarti Gupta, 'Transparency under Scrutiny: Information Disclosure in Global Environmental Governance' (2008) 8 *Global Environmental Politics* 1.

[57] Auld and Gulbrandsen, pg. 100.

processes improves accountability because the disclosed information enables regulated actors and the public to ask relevant questions and seek answers.[58] Secondly, by appealing to shared norms of openness, procedural transparency can enhance the acceptability of the certification scheme and thereby increase its legitimacy.[59] Finally, transparency can help convince participants that decision-making is conducted in a transparent and fair manner, thereby facilitating buy-in from more participants.[60] I will illustrate these points by reference to the GHG Protocol for Cities. As mentioned in earlier chapters, the various stakeholders involved in developing the GHG Protocol for Cities took pains to be as transparent in their deliberations and decision-making processes as possible. There were meetings open to the public and extensive consultation with cities that were not officially members of the working group but were likely to be users of the GHG Protocol. The information that went into the development of the GHG Protocol was made available on the Internet, accompanied by glossaries and user guides to make the information as accessible to the non-expert as possible. This is important, as the attributes of disclosed information are often central to the success of transparency initiatives. These attributes include whether the disclosed information is accessible, comprehensive, comparable, or relevant.[61]

When it comes to outcome transparency, the argument is that the disclosure of information about the environmental performance of cities to the public and stakeholders can enhance accountability because civil society groups can hold cities to account for their practices and performance. As Meidinger puts it, '[i]f a significant amount of information about a given practice is publicly available, then that practice becomes potentially accountable to a broad set of actors and values, at least in that it is subject to their criticism'.[62] Thus, transparency can play a role in enhancing what Keohane has termed 'external accountability', which refers to 'accountability to people outside the acting entity, whose lives are affected by it'.[63] The *raison d'être* of certification schemes such as the Compact of Mayors is to promote output transparency as a means of holding cities accountable to their commitments. It is too early to tell what the effects of transparency are for cities addressing climate change, but it can be said that urban climate law is characterized by a strong commitment to transparency as a mode of governance. Given the positive correlation between transparency and accountability, it can be argued that cities and their norms, practices, and voluntary standards contribute positively towards enhancing the accountability of the transnational climate change regime complex.

58 Ibid. 59 Ibid. 60 Ibid.

61 Klaus Dingwerth and Margot Eichinger, 'Tamed Transparency: How Information Disclosure under the Global Reporting Initiative Fails to Empower' (2010) 10 *Global Environmental Politics* 74.

62 Errol Meidinger, 'The Administrative Law of Global Private-Public Regulation: The Case of Forestry' (2006) 17 *European Journal of International Law* 47, pg. 82.

63 Robert O. Keohane, 'Global Governance and Democratic Accountability' in David Held and Mathias Koenig-Archibugi (eds), *Taming Globalization: Frontiers of Governance* (Polity Press 2003), pg. 141.

7.2.5 Sustainability

Sustainability of a regime complex is concerned with its resilience to external shocks and its long-term existence. In Keohane and Victor's view, sustainability is an important criterion because actors need long-term certainty about the rules of the game, so to speak. In the case of climate change, long-term certainty about rules plays a particularly pertinent role in achieving the goals of climate mitigation and adaptation because investors, both public and private, will not be willing to spend on long-term mitigation and adaptation strategies in the absence of clear signals about the long-term nature of rules mandating decarbonization and climate adaptation.

A good example to illustrate the importance of sustainability and resilience of a regime is the European Union Emissions Trading Scheme (EU ETS). The volatility of carbon prices within the EU ETS in the past few years was partly triggered by the economic crisis in the EU during the period of 2008 to 2009, but price volatility was also taken to indicate the existence of more structural problems with the EU ETS that threatened the stability and sustainability of the regional GHG emissions control scheme.[64] The EU institutions and its member states have also been taking a long time to reach agreement on a proposal to revise the EU ETS for the period of 2021–2030, further deepening pessimism about the future of the EU ETS.[65] Today, due to a confluence of factors including uncertainty about the future of the EU ETS and low oil prices, the price of an EU ETS allowance continues to be too low to create real incentives for industries to reduce GHG emissions and invest in low-carbon technologies.[66] The fall in EU ETS allowance prices has also created unanticipated challenges for the deployment of carbon capture and storage (CCS) technologies. The Rotterdam Capture and Storage Demonstration Project (ROAD) is a case in point.[67] ROAD is a joint venture project that is intended to capture and store 25 per cent of its annual 3 million tonnes of carbon dioxide emissions beneath the North Sea.[68] This 250 megawatt CCS plant was estimated to cost almost

[64] See, for example, Regina Betz, *What Is Driving Price Volatility in the EU ETS?* (Australasian Emissions Trading Forum, October/November 2006), online: www.ceem.unsw.edu.au/sites/default/files/uploads/publications/PagesfromAETFReviewOctNov06_web-2-1.pdf (accessed on 1 July 2016).

[65] For a status update of the EU ETS reform process and related documents, see European Council/Council of the European Union, 'Reform of the EU Emissions Trading Scheme', online: www.consilium.europa.eu/en/policies/climate-change/reform-eu-ets/ (accessed on 1 July 2016).

[66] Ibid.; also see Sara Stafanini, 'Climate Targets Suffer as Carbon Price Slumps' *Politico* (26 January 2016); research on the impact of the Swedish carbon dioxide tax and the EU ETS on productivity in the Swedish pulp and paper industry has shown that the carbon prices have been too low to create incentives for technological development; Tommy Lundgren et al., *Carbon Prices and Incentives for Technological Development* (Centre for Environmental and Resource Economics, Sweden, Working Paper, 2013: 4, 2013), online: www.cere.se/documents/wp/2013/CERE_WP2013-4.pdf (accessed on 1 July 2016).

[67] Rotterdam Capture and Storage Demonstration Project, online: http://road2020.nl/en/ (accessed on 1 July 2016).

[68] Ibid.

600 million euros.[69] It was to be funded by the European Union, the Dutch government, the port of Rotterdam, and indirectly by the EU ETS. In 2009, the project developers assumed that the price of an allowance (equivalent to one tonne of carbon dioxide) would rise from fifteen euros to thirty.[70] However, this assumption proved to be grossly incorrect, with the current price of an allowance hovering at seven euros. This created a funding deficit of more than 100 million euros, rendering the ROAD project non-viable and therefore put on hold for years.[71] The project was eventually completed, and at the time of writing, the ROAD project only operates at a fraction of its capacity, equivalent to the amount of subsidies and compensation it receives to cover operating costs, which exceed the costs of compliance with the EU ETS.[72]

While the EU ETS is a highly complex regulatory system and is prone to economic shocks because it is a market-based mechanism, its sustainability could have been better secured if the architects of the scheme had included components to withstand shocks. An example of such a component is a market stability reserve, which would allow the EU ETS regulator to maintain price stability by addressing fluctuations in the supply and demand of allowances.[73] During periods when there is a surplus of allowances in the EU ETS, the regulator could withdraw allowances and transfer them to the market reserve, and do the reverse during periods of scarcity of allowances. A market stability reserve has been approved as part of the EU ETS reform package and will start operating in January 2019.[74] This will eventually improve the resilience of the EU ETS, but until then, the EU ETS regime suffers a sustainability deficit in this respect.

Bringing the discussion back to whether cities and their norms, practices, and voluntary standards contribute towards building resilience and sustainability of the transnational climate change regime complex, it is relevant to recall the discussion in Chapter 6 on the reinforcing nature of the complementarities between the UNFCCC regime and urban climate law. Many key aspects of urban climate law have been designed to complement the norms and practices of the UNFCCC, while bypassing the issues that have caused gridlock in the international negotiations. It is relevant that C40 declared its support of the CBDRRC principle because this reinforces a core pillar of the international climate change institutional framework, thereby making the regime complex more resilient.

The commitment that cities have demonstrated towards putting in place a formal transparency framework based on the monitoring, reporting, and verification of their

[69] Tseard Zoethout, 'Closing Modern Coal-Fired Power Plants with CCS Will Slow Down Energy Transition' *European Energy Review* (11 December 2015), online: www.europeanenergyreview.eu/closing-modern-coal-fired-power-plants-will-slow-down-energy-transition/ (accessed on 1 July 2016).

[70] Ibid. [71] Ibid. [72] Interview No. 13.

[73] For discussion, see for example, Karsten Neuhoff et al., *Is a Market Stability Reserve Likely to Improve the Functioning of the EU ETS? Evidence from a Model Comparison Exercise* (Climate Strategies 2015).

[74] European Council/Council of the European Union, 'Reform of the EU Emissions Trading Scheme'.

emissions similar to the international transparency framework under the Paris Agreement is also noteworthy from a sustainability point of view. By investing significant amounts of resources and effort to implement a city-level transparency framework, cities are reinforcing the message of the Paris Agreement to all stakeholders that data-driven transparency and accountability will be a key platform of the future climate regime. This reduces uncertainty about the future rules and, it can be argued, will reinforce the signal to companies, industries, and governments to make the necessary policies and investments.

Furthermore, the Paris Agreement marks a milestone in the development of the climate change regime complex, but much remains to be done to operationalize its mechanisms and rules in the years to come. For example, the modalities and procedures regarding the NDCs have yet to be decided upon.[75] A lot of work will also need to be done prior to the first global stock-take, which will take place every five years starting before 2020.[76] In short, it is still early days, as work is under way to build the institutions needed to implement the Paris Agreement. Through voluntary schemes like the Compact of Mayors as well as ambitious policies and programmes within their jurisdictions, cities have collectively created a substantial amount of climate action at the sub-national level, which will go towards supporting the Paris Agreement in its early stage and the overall sustainability of the regime complex.

7.2.6 *Fairness*

Finally, Keohane and Victor include fairness as one of the evaluative criteria but do not say much about it except that multilateral institutions should not be evaluated on the basis of whether they achieve the 'utopian objective' of 'reflect[ing] abstract normative standards of fairness'.[77] Instead, 'fairness' simply refers to the notion that benefits ought to be distributed widely, and states that are willing to cooperate ought not to be discriminated against.[78] The imposition in the Kyoto Protocol of binding GHG emission targets on developed countries, and not on developing countries, was used to illustrate the fairness point. In Keohane and Victor's view, 'the absence of binding rules for some states was of questionable fairness'.[79]

Fairness can be said to comprise two key aspects: distributive justice and procedural fairness. When considering whether fairness is achieved in a particular context, the question of distributive justice would be whether regime complexes 'create solutions and systems which take into account society's answers to [the] moral issues

[75] UNFCCC, Ad Hoc Working Group on the Paris Agreement (APA), online: http://unfccc.int/bodies/apa/body/9399.php (accessed on 5 August 2016).

[76] For discussion, see Eliza Northrop, Cynthia Elliott, and Melisa Krnjaic, *4 Key Questions for the Design of the Global Stocktake* (19 May 2016), online: www.wri.org/blog/2016/05/insider-4-key-questions-design-global-stocktake (accessed on 5 August 2016).

[77] Keohane and Victor, pg. 17. [78] Ibid. [79] Ibid.

of distributive justice'.[80] Procedural fairness, on the other hand, is concerned with 'what the participants perceive as right process'.[81] I adopt Naiki's argument that the criterion of procedural fairness 'is difficult to apply in the context of regime complexes, because the question is whether a regime complex is made under the right process. Yet, regime complexes often emerge without the right process or order, and that is why a comprehensive regime was not yielded.'[82] She suggests that a possible line of enquiry is whether each regime (within the regime complex) is fair in the procedural sense – for instance, whether a voluntary certification scheme can be said to have achieved input or process legitimacy.[83]

It can be argued that the norms, practices, and voluntary standards that cities have developed and implemented have been able to sidestep questions of distributive justice in ways that the UNFCCC regime cannot because the latter is intended to be an international agreement that secures universal participation. An international agreement that fails to address developing country needs and interests will be perceived to be an unfair agreement that will be rejected by developing countries that make up the majority of the international community. Consequently, the UNFCCC regime has had to deal with the issues of loss and damage, climate finance, and promoting technology transfer to developing countries, to name a few key examples. Urban climate law, in contrast, involves a select group of global cities that have voluntarily signed up to a set of norms, practices, and voluntary standards. As obligations are not imposed upon cities and cities are persuaded that urban climate law only brings positive benefits such as reputational gain and improved access to sources of climate finance, questions of distributive justice tend not to arise. There is therefore limited opportunity for cities to contribute towards addressing critical issues of distributive justice that, for some, lie at the heart of the climate change debate. As for procedural fairness, we have already considered this when evaluating the related issues of procedural transparency and accountability, and thus the arguments will not be repeated here. As concluded earlier, it can be argued that urban climate law scores relatively well on procedural fairness because of its contribution towards enhancing the transparency and accountability of the regime complex.

To summarize the discussion, the question that this part of the chapter sought to consider is whether the normative activities of global cities make a positive contribution towards the performance of the transnational climate change regime complex. First, in relation to coherence – which refers to the consistency and compatibility of the practices and outputs of the various regimes within a regime complex – urban climate law can be said to be contributing positively towards the coherence of the transnational climate change regime complex. What is particularly noteworthy is the alignment between urban climate law and the norms and practices

80 Thomas M. Franck, *Fairness in International Law and Institutions* (Oxford University Press 1998), pg. 8.
81 Ibid., pg. 7.
82 Naiki, pg. 156.
83 Ibid.

of other regimes, streamlined monitoring and reporting obligations that build synergies between various regimes and urban climate law, and the extent to which global cities contribute towards learning processes within the regime complex.

Secondly, on the textual determinacy of rules, the voluntary standards that global cities have developed and implemented can be said to contribute positively towards determinacy in the regime complex because of their high level of textual specificity and clarity in normative content. Thirdly, to the extent that epistemic quality refers to the consistency between urban climate law and IPCC science, urban climate law enjoys a respectable level of epistemic quality. At the same time, global cities and their networks do not rely only on scientific knowledge but also consult the expertise of urban planners, architects, financiers, entrepreneurs, and development consultants, to name a few categories of experts, in developing and implementing their practices and voluntary standards. This increases the epistemic quality of urban climate law. Fourthly, considering that transparency is an important aspect of accountability, urban climate law contributes positively towards improving the accountability of the regime complex because its dominant mode of governance is underpinned by transparency mechanisms. On the fifth criterion of sustainability, of note are the complementarities between the UNFCCC regime and urban climate law and how the substantial amount of urban climate action goes towards supporting the Paris Agreement in its early stage and enhancing the overall resilience of the regime complex. It is on the final criterion of fairness that urban climate law makes little, if any, positive contribution because the nature of urban climate law is such that it presents cities with limited opportunities to address distributive justice issues.

7.3 REFLECTING ON THE SIGNIFICANCE OF CITIES FOR THE STUDY OF INTERNATIONAL LAW

Section 7.2 sought to evaluate if global cities make a negative or positive normative contribution towards the performance of the transnational climate change regime complex and concluded that it was the latter case. If so, there is a good argument to be made that contemporary global governance processes in other areas such as healthcare and tackling gender inequality ought to encourage greater participation of global cities and further harness the governance potential of cities and their transnational networks. However, before making such recommendations, it may be apposite to pause and consider some questions that the emergence of cities as transnational lawmaking actors poses for international law and international relations more broadly. It is timely to embark on such contemplation because, for a long time, international law has had highly limited engagement with the lively, multidisciplinary debate concerning the growing prominence of cities in international affairs. As this book has sought to demonstrate, the rise of cities as transnational lawmakers raises considerations that are relevant to the study of international law, and international law scholars are encouraged not to continue neglecting cities in their research.

7.3.1 *Challenging Statist Conceptions of International Law*

This book may be considered post-Westphalian, as it seeks to move beyond the focus of the international legal system upon states as the primary subjects.[84] The starting point in this book for exploring the emergence of cities as lawmaking actors was adopting the New Haven School's approach to international law. There are no subjects or objects in the New Haven School – only participants – and participants 'include those formally endowed with decision competence, for example judges, and all those other actors who, though not endowed with formal competence, may nonetheless play important roles in influencing decisions'.[85] This recognition that many actors, apart from states, can play a role in international lawmaking processes forecloses the argument about whether cities are recognized to be a class of actors in international law and how recognition ought to be conferred.[86]

By positing that cities are participants in the international legal order from the outset, my focus has been on examining their norm-setting practices and explaining their significance in terms of how they steer the behaviour of cities towards climate change mitigation and adaptation. Further, Myres McDougal and Harold Lasswell, pioneers of the New Haven School approach, have defined law as 'a process of authoritative decision by which the members of a community clarify and secure their common interests'.[87] According to this definition of law, the norms and voluntary standards that global cities have developed and subscribed to would count as international lawmaking, as the participating governments are authorities making collective decisions on the regulation of GHGs which they have the power to implement in their jurisdictions. The governments of global cities also articulate their common interests by signing up to declarations and other public statements issued at international conferences, for example.

84 I use the term 'Westphalian' to refer to an international system that is underpinned by the central notions of states as the primary subjects and objects of international law and wherein international law is created on the basis of consent by equal and sovereign states; see James Crawford, *Brownlie's Principles of Public International Law* (Oxford University Press 2012), pgs. 287–288.

85 W. Michael Reisman, 'The View from the New Haven School of International Law' (1992) 86 *American Society of International Law Proceedings* 118, pg. 122.

86 One can also question if formal recognition really matters. Aust, for example, argues that 'the more cities become accepted as partners in global governance, the less relevant will their informal status (as opposed to states and international organizations) be in the long run'. What matters is that there is growing acceptance of cities by international actors and audiences, and Aust suggests that this is just beginning to happen for networks like C40. Finally, he concludes that cities constitute an emerging 'new class of actors in international law' even though it is difficult to 'pin [this] down in positivist/formalist terms'; Aust, pgs. 273–275.

87 Harold D. Lasswell and Myres S. McDougal, *Jurisprudence for a Free Society: Studies in Law, Science and Policy* (Martinus Nijhoff Publishers 1992), pg. xxi.

My response to questions such as whether the transnational governance activities of cities challenge our perspectives on actors and subjects of international law and whether they call into question the traditional distinction between law and non-law is that these questions do not really get us anywhere because our answers ultimately boil down to the jurisprudential view of the international legal order that each of us holds. Depending on one's jurisprudential leanings, the state is the only significant actor on the international stage or it is one of many actors, albeit a very important one, or it is simply one of many participants that has the resources and power to engage in international affairs.[88] Thus, I would suggest that the study of cities and their normative activities reinforces the idea that international law no longer refers solely or even primarily to the law that governs the rights and obligations of states in co-existence and that there are various theoretical conceptions of the international/transnational/global legal order. International legal scholarship is enriched by the recognition of these various schools of thought and a move away from state-centric versions of international lawmaking, which hinder the crafting of creative solutions to our global collective action problems such as climate change. By creative solutions, I mean solutions that recognize the importance of multilevel governance and enlist the participation of multiple governance actors, including cities, to tackle global governance challenges more effectively.

7.3.2 *Recognizing the Impact of International Law on Cities*

It should be noted that, while cities are on the margins of international law and are hardly recognized as worthy of scholarly analysis, international law already directly affects the world's cities by having determinative influences on decision-making processes and regulation at the sub-national level. Frug and Barron have examined how international institutions and laws are bringing the governments of cities within their regulatory reach.[89] They argue that international law is beginning to play a role in defining the relationship between cities and the states in which they are located: 'Parties negotiating international trade agreements, international tribunals arbitrating commercial disputes, United Nations' rapporteurs investigating compliance with human rights obligations, and international financial institutions formulating development policy have all begun to express interest in the legal relationship between cities and their national governments.'[90] Therefore, a new set of international rules and regulations governing cities has emerged, and Frug and Barron coin the term 'international local government law' to refer to it.

[88] For discussion of how different perspectives on international law might shape the narrative of the international legal significance of sub-national efforts in addressing climate change, see Hari M. Osofsky, 'Multiscalar Governance and Climate Change: Reflections on the Role of States and Cities at Copenhagen' (2010) 25 *Maryland Journal of International Law* 64, pgs. 75–79.

[89] Frug and Barron. [90] Ibid., pg. 1.

Frug and Barron illustrate the nature of international local government law with reference to decisions by international arbitration tribunals that have arisen from international investment disputes. One such example is the well-known decision in *Metalclad v. United Mexican States.*[91] Briefly, the facts of the case are as follows: Metalclad is a US company that had purchased a landfill in the Mexican municipality of Guadalcázar. The dispute started when Guadalcázar's municipal authorities denied Metalclad permission to construct and operate the landfill despite the company having received guarantees and reassurances from higher levels of government that it could proceed with the landfill project.[92] Domestic legal proceedings failed to determine this issue. Metalclad then sought to resolve the dispute by bringing a claim under the North American Free Trade Agreement (NAFTA). The NAFTA arbitration panel ruled in favour of the company. In the panel's opinion, Mexico's breach resulted from the central government's failure to stop the city of Guadalcázar from asserting its expansive view of its domestic legal authority.[93] Frug and Barron argue that, in the extensive literature on the arbitration panel's decision in *Metalclad*, what is often overlooked is the fact that the case turned on a dispute about the scope of city power in Mexico.[94] In their view, '*Metalclad* crafted a rule that limited the ability of cities to make their own interpretations of local regulatory authority, thus taking a position on a central issue of Mexican local government law.'[95] Thus, it can be argued that the concerns raised by the *Metalclad* decision go beyond the international legal system's capacity to extend the scope of investor immunity from governmental regulations.[96] The decision also demonstrates how the international legal system can restructure the relationship between cities and higher levels of government within a state.[97]

While international local government law focuses our attention on the ways in which international law can have an impact on local decision-making processes in cities and therefore shape the everyday urban reality, the deliberations in this book on urban climate law shed light on a different relationship between the world's cities and the international legal order. Urban climate law is evidence that cities can exert normative effect within the transnational legal order and that the relationship between cities and the international legal order does not always have to be mediated by the state. International local government law and urban climate law together provide a fuller picture of how international law has an impact on cities and vice versa. Importantly, the above discussion also illustrates the relevance of cities to the study of international law and buttresses the claim that it is timely for international legal scholarship to examine the emergence of cities as transnational governance actors and the implications of this development.

[91] *Metalclad Corporation v. United Mexican States* ICSID Case No. ARB(AF)/97/1; 40 I.L.M. 36 (2001).
[92] *Metalclad Corporation v. United Mexican States*, para. 50. [93] Ibid.
[94] Frug and Barron, pg. 41. [95] Frug and Barron, pg. 43. [96] Frug and Barron, pg. 44.
[97] Ibid.

7.3.3 *Critical Evaluation*

While this book has considered the emergence of cities as international actors to be a positive development, the rise of cities should not be praised uncritically without consideration of some of the less positive aspects. In this regard, international law doctrines and scholarship can provide a prism through which the transnational governance activities of cities can be considered in a critical manner.

A consequence of the emergence of cities as international actors is the challenge posed to the consent-based structure of international law.[98] If formal international lawmaking processes such as that of the UNFCCC are to include cities in negotiations and give cities voting rights, for example, how would the international system uphold the requirement of consent? Another related concern is that the inclusion of cities is likely to increase the complexity of international negotiations, which are already bogged down by the immense difficulties of reaching compromise on complex issues amongst states with disparate interests. There are a number of possibilities. One possibility would be to reform current international legal decision-making processes to give cities greater participation rights in the early stages of treaty negotiations. Hari Osofsky has suggested that such reforms can include creating forums in which city governments can meet directly with state representatives involved in negotiations or inserting provisions into formal agreements that acknowledge the role of cities in regulating climate change. As noted in Chapter 1, the Cancun Agreements recognized local and sub-national governments as 'governmental stakeholders' of the UNFCCC regime in 2010, so the latter recommendation has already been taken up.[99] These reforms, while they would create connections between urban efforts to tackle climate change and the UNFCCC regime, would not give cities a genuine voice in the negotiations that will continue to take place among states.

Another possibility would involve more significant reform that would allow cities and other actors apart from states to participate formally in international climate agreements that might supplement the Paris Agreement. One therefore may envision a suite of agreements under the UNFCCC umbrella, the Paris Agreement being one concluded by states and other agreements being concluded by sub-national governments, corporations, and industry associations, to name a few. The structuring of such agreements is likely to raise difficult logistical issues as well as normative concerns surrounding inclusion, fairness, and accountability. Further, the promulgation of such agreements within the UNFCCC framework would be considered revolutionary and perhaps even sacrilegious in the eyes of traditionalists who adhere to the Westphalian model of international law.

98 Aust, pg. 276.

99 Decision 1/CP.16 The Cancun Agreements: Outcome of the Work of the Ad Hoc Working Group on Long-term Cooperative Action under the Convention, para. 7.

I would argue that it is immaterial if scholars and practitioners choose to explore the benefits and disadvantages of various options and ultimately decide that the current approach to international lawmaking remains the best one. What is important is that such deliberations occur in the first place. These discussions ought to be encouraged, to reframe the discourse about the involvement of cities in transnational governance processes and to explore creative possibilities for alternative governance arrangements in the international system. This is not an academic point, as the international system is already witnessing attempts to formally institutionalize the role of cities in global governance. The Global Parliament of Mayors, which describes itself as 'a new global governance body of cities', convened its inaugural meeting in September 2016 at The Hague.[100] Its mission statement highlights many of the themes discussed in this book. First, the mission statement sets out that it is 'inconceivable that national and international bodies discuss and decide on policy actions without cities and their mayors present at the table', given that more than half of the world's population lives in cities.[101] In line with the trend of cities playing an increasingly active role in international affairs, 'the Global Parliament of Mayors claims the right not only to be involved, but to be (with others) agenda setters in supranational organizations'.[102] Specifically, mayors 'will cooperate on critical issues such as climate change, refugee crises, pandemic diseases, inequality and urban security'.[103] What distinguishes the Global Parliament of Mayors from transnational city networks like C40 is that the former aims to be an institution that represents the collective interests of cities all over the world and participate in global decision-making processes on equal footing with states and international organizations. In other words, it seeks to break the monopoly that national governments have on international decision-making processes. If the Global Parliament of Mayors is to issue recommendations, resolutions, and decisions, this would raise interesting questions of how the international law made by cities will interact with international law made by states, and if the former will be treated differently and in what ways.

Finally, the rise of global cities and the increasing direct collaboration between cities and international organizations are developments that may raise concerns about the possible disintegration of the state, fragmentation of the international legal order, and the negative effects of decentralization (for example, its contribution to inequality within global cities). The third issue, while very interesting, is one that urban sociologists and economists are better placed to address.[104] The first

[100] Global Parliament of Mayors, *Press Release: Inaugural Meeting of Global Parliament of Mayors in The Hague, Netherlands*, The Hague/New York, 1 September 2016, online: www.globalparliamentofmayors.org/press (accessed on 25 November 2016).

[101] Global Parliament of Mayors, 'Mission Statement', online: www.globalparliamentofmayors.org/mission-statement (accessed on 25 November 2016).

[102] Ibid. [103] Ibid.

[104] See, for example, Remy Prud'homme, 'The Dangers of Decentralization' (1995) 10(2) *World Bank Research Observer* 201; Yener Altunbas and John Thornton, 'Fiscal Decentralization and Governance' (2012) 40(1) *Public Finance Review* 66; Local Development LLC, *The Role of*

and second issues are likely to be of interest to international legal scholars, and it is on these two issues that some views will be offered here.

In this book, I have sought to show that the interesting question is not whether the rise of global cities contributes to the disintegration of the state, but rather what the rise of global cities tells us about the changing nature of the state and its ability to take on different structural forms on different occasions and for different purposes. In short, the rise of global cities provides fodder for the debate on the nature of the modern state, and we ought to contest the simplistic notion that the rise of sub-national actors threatens the integrity of the state as a unitary entity. I would argue that the state is far from being a unitary entity at all times, and the recognition of its mutable nature creates interesting possibilities for transnational governance arrangements.

While states have struggled to develop a comprehensive climate change treaty and the resulting international governance arrangement is made up of loosely connected institutions, transnational networks made up of non-state actors and sub-state actors can play a meaningful role in supplementing state-led governance. The emergence of transnational city networks has not threatened to supplant states, as the rhetoric might portray, but in a more subtle way strengthens and reinforces the transnational climate change governance order. Furthermore, it should be noted that states facilitate the transnational forces that enable cities to engage in global governance. States are not being ambushed and taken by surprise by the increasingly active role of global cities in international affairs. Instead, it can be argued that national governments have been convinced that cities can play a meaningful role in global governance, particularly in issue areas which require localized action – such as tackling infectious diseases – and are creating and supporting institutional platforms to enlist the involvement of cities alongside other actors. Therefore, when the UN-Habitat and the UNFCCC create partnerships and programme platforms to work directly with cities, for example, states often also participate as partners in these programmes and multi-stakeholder partnerships. In 2013, ICLEI succeeded in enlisting the support of national governments for the creation of frameworks to promote cooperation on urban climate adaptation and mitigation activities. The 'Friends of Cities' group includes France, Germany, Indonesia, Mexico, the Netherlands, Peru, and South Africa.[105] The South African government supported the adoption of the Durban Adaptation Charter for Local Governments in 2011 at the UNFCCC Conference

Decentralisation/Devolution in Improving Development Outcomes at the Local Level: Review of the Literature and Selected Cases [Prepared for the United Kingdom Department for International Development (South Asia Research Hub), November 2013], online: www.delog.org/cms/upload/pdf/DFID_LDI_Decentralization_Outcomes_Final.pdf (accessed on 26 November 2016).

105 ICLEI, 'Friends of Cities: Multi-level Partnerships for Scaling Up Local Action', online: www.iclei.org/climate-roadmap/pressroom/news/news-details/article/friends-of-cities-multi-level-partnerships-for-scaling-up-local-action.html (accessed on 1 August 2017).

of the Parties in Durban.[106] In 2013, the French government supported the Nantes Declaration of Mayors and Subnational Leaders on Climate Change, which commits city authorities to work with national governments and non-state actors to address climate change.[107]

States also sponsor many of the capacity-building programmes for cities to act on climate change and other global governance challenges. For example, the funders of the Cities Development Initiative for Asia include the governments of Germany, Sweden, and Austria.[108] There are, of course, also counter-examples of national governments that have stood in the way of their cities exercising greater independence.[109] Anecdotal evidence suggests that domestic politics weigh heavily in decisions to hold city governments back, rather than concerns about the integrity of the state as a political entity.[110] The disaggregation of the state into its component parts is a contemporary reality, and I would argue that it is a source of flexibility and resilience rather than a sign of weakness and fragmentation. The ability of the state to disaggregate into component parts which seek to create networks and partnerships with their counterparts abroad (e.g. transnational networks of judges, parliamentarians, and regulators of the financial services industry) is evidence of the ability of the state to mutate and reconfigure to address different challenges. This flexibility not only benefits the state but also the international system as a whole, because global governance has become increasingly complex, and the one-size-fits-all solution offered by traditional treaty-based regulation by international organizations is no longer adequate. The challenge for international law is finding ways to maintain accountability, fairness, and legitimacy in the international community as transnational lawmaking and implementation processes become more informal and inclusive of various participants.

Finally, it is important to maintain some critical perspective about the role that cities can play in global governance. This book has sought to shed light on recent developments in transnational climate change governance that show that cities are playing an increasingly active role in international affairs. It has also sought to uncover the normative effects of cities' governance activities. These are novel developments and, in light of the slow progress in international climate change negotiations up till the Paris Agreement was concluded in December 2015, are a source of hope for the future of global climate action. However, this book does

106 ICLEI, 'Friends of Cities: Good Practices in Multi-Level Partnerships on Scaling-Up Climate Action', online: www.iclei.org/fileadmin/PUBLICATIONS/Papers/FriendsofCities_march2015_for_CITEGO.pdf (accessed on 1 August 2017), pg. 7.

107 Ibid., pg. 3.

108 Cities Development Initiative for Asia, 'Funding and Implementation', online: http://cdia.asia/who-we-are/funding-and-implementation/ (accessed on 26 November 2016).

109 Sam Barnard, *Climate Finance for Cities: How Can International Climate Funds Best Support Low-Carbon and Climate Resilient Urban Development?* (Overseas Development Institute, Working Paper 419, 2015), pg. 20.

110 Interviews Nos. 2, 4, and 5.

not argue that cities offer a panacea for complex world governance issues while the inter-state system has failed to provide adequate solutions. It must be recalled that cities are also domestic political entities. Notwithstanding their globalist aspirations, they are able to pursue global governance ambitions only if they have the financial capabilities to do so – and the ability of cities to levy and collect taxes is a question of national politics. Further, however intertwined a city may be in the global economy, a global city is still enmeshed in a national constitutional structure which dictates the structure and size of a city's government. National law and politics also dictate the kind of powers that a city government has. Hence, it is important to retain some perspective on how much of a challenge global cities pose to states and the international legal order. It is helpful to consider the rise of cities less in terms of challenging the status quo and more in terms of alternatives to state-centrism when we consider the international legal order.

7.4 CONCLUSION

This chapter sought to normatively assess the contribution of urban climate law to the overall performance of the transnational climate change regime complex using six evaluative criteria proposed by Keohane and Victor: coherence, accountability, determinacy of rules, sustainability, epistemic quality, and fairness. Keohane and Victor did not say much about each criterion, leaving room for future elaboration. Thus, in carrying out the normative assessment in this chapter, I also sought to flesh out the criteria and, in some cases, provide illustrative examples from other areas of climate law and policy.

The evaluation highlights some strengths and weaknesses of urban climate law. A key strength of urban climate law is its commitment towards transparency. By making information concerning their climate actions available to the public and therefore subjecting themselves to public scrutiny, cities hold themselves accountable and in the process enhance the accountability of the regime complex. Where urban climate law does not fare as well is on the issue of fairness, specifically on questions of distributive justice. By design, urban climate law has reframed the climate change debate and managed to avoid the contentious issues in the international climate change negotiations such as binding GHG emission reduction targets and firm financial commitments on the part of developed countries to assist developing countries in their climate mitigation and adaptation efforts. However, these contested issues are also the ones that raise difficult questions of distributive justice. Without having to deal with these matters, urban climate law has created a substantial amount of climate action. This goes towards supporting the Paris Agreement in its early stages and enhancing the overall sustainability of the regime complex, but it does not contribute towards answering issues of distributive justice and fairness in the regime complex.

Finally, the chapter turned towards a set of questions concerning the rise of cities in international affairs. These broad questions are difficult and complex. All that this chapter has sought to do is to draw together some salient observations in response. What is clear is that international law scholars should not ignore the rise of cities. If anything at all, the rise of cities points to interesting possibilities for the future of international law as international law scholars recognize the mutable nature of the modern state and the governance options that are made available by the disaggregation of the state.

8

Conclusion

8.1 INTRODUCTION

The UNFCCC entered into force in 1994, the first milestone in the international community's search for a collective response to climate change. The attempt to achieve consensus amongst the 197 countries that ratified the UNFCCC on what needed to be done next to achieve the treaty's objectives was always going to be difficult. Social cooperation has been key to the survival and success of our species, and we *Homo sapiens* have created many myths and fictions that have given us unprecedented ability to cooperate flexibly in large numbers. Wolves and chimpanzees cooperate in groups, but they can do so only with a limited number of other individuals with whom they share intimate social bonds. It has been suggested that *Homo sapiens* managed to cross the critical threshold and eventually create cities and empires of millions of people because of fiction. 'Any large-scale human cooperation – whether a modern state, a medieval church, an ancient city or an archaic tribe – is rooted in common myths that exist only in people's collective imagination.'[1]

The state is a myth, and a powerful one. Two Singaporeans who have never met one another might risk their lives to save each other because they believe in the existence of the state of Singapore, its history as a modern nation, and a common national identity. This myth, however, can also stand in the way of collective action. One of the most contentious issues that marred the international climate change negotiations for years can easily be summarized thus: Why should my country reduce GHG emissions when yours is not required to? Why should my country bear the risk of slower economic growth while yours is allowed to burn unlimited amounts of fossil fuels to power factories, ports, and homes? *Homo sapiens* have single-handedly caused anthropogenic climate change within the equivalent of a fraction of a second in Earth's cosmic history, and our myths have hindered us from effectively working together to remedy the problem.

[1] Yuval Noah Harari, *Sapiens: A Brief History of Humankind* (Vintage Books 2011), pg. 30.

However, perhaps not all is lost. The lack of progress in the international climate change negotiations has inspired other actors to find solutions. Universities have been working together with venture capitalists and philanthropic organizations to develop clean technologies for the future.[2] Engineers are working with physicists to explore the potential for airborne wind energy, a renewable energy technology that uses airborne devices (kites in particular) to harness wind power.[3] Architects and urban planners are increasingly incorporating green roofing and other design features to cool buildings and reduce the urban heat effect in cities.[4] Some financial regulatory agencies now require companies to disclose their climate change risks.[5] Regions and states (within a federal system) have created voluntary GHG emissions trading programmes.[6] The list goes on. In the midst of this proliferation of climate change responses, cities have also crafted their unique response.

This book has examined the emergence of cities as climate change governance actors, particularly their contribution towards generating norms, practices, and voluntary standards that aim to steer cities towards GHG emissions reductions and developing low-carbon development pathways for the future. Cities have transcended mythical state boundaries to create cooperative networks that allow them to more effectively learn from one another, share best practices, and diffuse ideas about how a city ought to be responding to climate change. This final chapter will conclude the research over four additional sections. Section 8.2 will provide an overview of the issues addressed by this book. It will provide a summary of the research and answer the five research questions posed in Chapter 1. Section 8.3 will set out some of the implications of this research for environmental law scholars, international legal scholars, and practitioners. Section 8.4 suggests avenues for further research. Section 8.5 will offer some final remarks.

[2] However, the clean technology bubble has burst, leading to a significant slowdown in venture capital investment in clean energy technology start-ups since 2009; Benjamin Gaddy and Varun Sivaram, 'Clean Energy Technology Investors Need Fresh Support after VC Losses' *Financial Times* (26 July 2016).

[3] Uwe Ahrens, Moritz Diehl, and Roland Schmehl (eds), *Airborne Wind Energy* (Springer 2013).

[4] Sarah Murray, 'Nature Is Now a Weapon against Threat of Global Warming' *Financial Times* (1 June 2016).

[5] An example is the US Securities and Exchange Commission, though it has recently been criticized for its lax enforcement of climate risk disclosure rules; David Gelles, 'S.E.C. Is Criticized for Lax Enforcement of Climate Risk Disclosure' *New York Times* (23 January 2016).

[6] The Western Climate Initiative (WCI) and the Regional Greenhouse Gas Initiative (RGGI) are two examples. WCI is a non-profit corporation that provides technical assistance to support the implementation of state and provincial GHG emissions trading programmes. In 2014, California and Quebec linked their cap and trade programmes under the auspices of WCI. More information is available at www.wci-inc.org. RGGI is the first cap and trade programme in the United States to reduce carbon dioxide emissions from the electricity generation sector. Its members are the states of Connecticut, Delaware, Maine, Maryland, Massachusetts, New Hampshire, New York, Rhode Island, and Vermont. More information is available at www.rggi.org (both websites accessed on 15 August 2016).

8.2 THE RESEARCH FINDINGS

In this section, we return to the five research questions that were set out in Chapter 1.

1. *What recent developments suggest the rise of cities in international affairs?*

Early indications in the research for this book suggested that the emergence of cities as global governance actors in the climate change context was part of a broader trend in international affairs. Chapter 3 is devoted to answering this question, thereby also providing the wider context for understanding the rise of cities as lawmaking actors in the transnational climate change regime complex. This chapter also sought to provide an account of the involvement of cities in international affairs, to ground the argument that contemporary developments suggest that it is time for international legal scholars to pay more attention to cities, which have traditionally been neglected because of state-centric conceptions of international lawmaking.

Chapter 3 contained discussion of four broad categories of activity that exemplify how cities are participating in transnational legal and political processes. What is unique about their involvement is that the cities have adopted practices and policy positions independent of or contrary to that of their states in many instances. The four categories are these: (1) a city implementing international law on its own accord when its national government is reluctant or refuses to do so, (2) city diplomacy, (3) cities developing their independent local and transnational policies and strategies to manage global risks such as terrorism, (4) cities forming organizations to represent urban interests in international forums and/or to pursue governance objectives. Two examples of the first category of actions canvassed in the chapter are the adaptation and implementation by American cities of the Kyoto Protocol and CEDAW, two international treaties that the United States has not ratified. As for city diplomacy, there are many instances of bilateral interactions between cities in different states. One of the examples in this chapter was the mayor of Tokyo's official visit to his counterpart in Beijing during a period of great Sino-Japanese tensions, with the explicit goal of using city diplomacy to defuse the increasing political tension at the interstate level.

2. *What have cities been doing to govern climate change, and which of these governance activities generate normative effects transnationally?*

Over 2,000 cities have developed climate change action plans to reduce their GHG emissions and adapt to the impacts of climate change. In this book, I have chosen to focus on a sample of global cities to illustrate the local-global connections in urban climate actions. The global cities I have focused on are

Rotterdam, London, New York City, Seoul and Mexico City. Each of these cities faces different challenges and opportunities in addressing climate change. For example, Rotterdam has high GHG emissions because of industrial activities and its port. The city has therefore focused on reducing port-related GHG emissions, developing infrastructure that connects various parts of the city such that heat generated by the power plant can be channelled towards greenhouses and homes. As a delta city with exposure to the North Sea, the city's government has taken water adaptation very seriously and developed significant expertise.

The governance activities that global cities have undertaken generate transnational normative effect when they create cross-border networks to facilitate the sharing of norms, practices, and voluntary standards. While there are a number of city networks that address sustainability and environmental issues including climate change, only one network is made up of large global cities and is solely focused on tackling climate change. This network, C40, is therefore the subject of detailed discussion in Chapter 5. Setting up the network infrastructure to facilitate the sharing of norms and practices is not enough. What is noteworthy is that cities have developed forms of voluntary certification (such as the Compact of Mayors), standardized accounting protocols that cities can use to measure and report their GHG emissions, and publicly accessible systems of data disclosure so that they can be held accountable for their commitments. These voluntary certification schemes, protocols and data disclosure systems are developed in partnership with international organizations, global civil society actors, states and private-sector consultancies, and are transnational in nature.

3. *How do the norms, practices, and voluntary standards generated by cities and transmitted by their networks lead to cities reducing their GHG emissions and increasing their climate resilience?*

This book argues that there are two key processes that play a critical role in linking urban climate law with changing actual practices on the ground. These pathways involve the promotion of reflexivity and norm diffusion. Reflexivity involves actors receiving new information and adjusting their practices accordingly to achieve desired objectives such as increased productivity at the workplace or improved energy efficiency. In this line of thinking, voluntary standards are regulatory tools that uncover new information for cities and also provide a systematic approach that cities can use to foster self-reflection and new thinking.

The literature has identified many norm diffusion mechanisms, but most of them can be grouped into the following categories: competition, learning, and imitation. This book argues that competition plays a negligible role in shaping how norm-setting practices are diffused across cities, but learning and imitation offer compelling explanations.

4. *How do the norms, practices, and voluntary standards generated by cities and transmitted by their networks relate to those of the UNFCCC regime?*

While some commentators claim that the UNFCCC regime is no longer at the core of the transnational climate change regime complex, this book argues otherwise. The UNFCCC remains an important central pillar of international cooperation on climate change and provides the necessary institutional and normative frameworks. While it does not have formal authority over other transnational governance initiatives, particularly those that are convened by non-state actors, it is often the case that these initiatives rely on the UNFCCC regime for normative guidance and to set the global agenda on climate change responses. In order to ensure that the transnational climate change regime complex performs well – that is, steers various communities and actors towards climate change mitigation and adaptation – it is important to ensure that the individual regimes that make up the regime complex do not conflict.

A central argument in this book is that urban climate law not only does not come into conflict with the norms and practices of the UNFCCC regime, but the former has also been designed in ways that complement and reinforce the latter. For example, it is noteworthy that the Compact of Mayors is designed to be the city-level equivalent of the transparency framework created by the Paris Agreement. During the run-up to the COP in Paris, the Compact of Mayors sought to have city-level initiatives complement and support the international climate negotiations by adopting the pledge-and-review approach that underpins the Paris Agreement. Subsequently, the 'City Climate Commitments'[7] that cities are required to pledge upon joining the Compact parallel the NDCs that states are required to submit pursuant to the Paris Agreement. A detailed analysis of how urban climate law and the UNFCCC regime are linked in strategic and deliberate ways is found in Chapter 6.

5. *Do global cities make a positive normative contribution to the global constellation of climate change governance activities?*

It is not possible to answer this question by referring to whether cities have achieved the goals and targets they have set themselves because we are still at the early stages of promoting the uptake of urban climate law and putting it into action. It is simply too early for significant results. However, we can ask ourselves the question of whether urban climate law contributes positively to the overall performance of transnational climate change governance by evaluating it in accordance with well-established normative criteria. Chapter 7 concluded that, from a bird's-eye view, global cities contribute positively in normative terms to the performance of the transnational climate change regime complex. While urban climate law does not sufficiently

[7] Compact of Mayors, online: www.compactofmayors.org/resources/ (accessed on 1 July 2016).

engage with issues of substantive justice, it does contribute to the regime complex's coherence, accountability, and sustainability.

8.3 IMPLICATIONS OF THE RESEARCH

8.3.1 *Contribution to Existing Literature*

This book contributes to a number of different strands of literature. First, this research contributes a legal dimension to the large body of literature on cities and climate governance. This body of literature has tended to focus on cities within a multilevel governance system such as the European Union, on producing single case studies on cities or on comparative analysis of a few cities. At the time of writing, this book is the first systematic study to consider the role of cities as lawmakers in transnational climate change governance. At the same time, I seek to contribute towards better understanding of the governance capabilities of cities.

The research in this book also contributes to the global city literature. Research on global cities has tended to focus on their networked nature in the global economy and global city competitiveness, which is concerned with global trade, financial flows and investment, and the quality of life that each global city offers. In recent times, scholars like Michele Acuto have forged a new direction in the research on global cities by focusing on their role in international affairs.[8] In *Global Cities and Climate Change*, Taedong Lee coins the term 'trans-local relations' to refer to the involvement of cities in international affairs and environmental governance.[9] His work centres on uncovering and explaining what shapes the participation of cities in transnational climate networks and how involvement in these networks shapes the agency capacity of cities. Another example is Sofie Bouteligier's *Cities, Networks and Global Environmental Governance*.[10] In her book, Bouteligier studies two transnational city networks – Metropolis and C40 – and analyzes how global cities perform the role of strategic sites of global environmental governance by serving as hubs that concentrate knowledge, institutions, and infrastructure. The research in this book has benefitted greatly from this strand of the global city literature and in turn contributes to it by exploring the hitherto unexamined role of global cities as norm-generative actors.

This book also contributes to the academic discourse on transgovernmental networks and the disaggregation of the modern state. In A *New World Order*, Anne Marie Slaughter examines the emergence of sub-state networks as an important

[8] Michele Acuto, *Global Cities, Governance and Diplomacy: The Urban Link* (Routledge 2013); Michele Acuto, 'Global Cities: Gorillas in Our Midst' (2010) 35 *Alternatives: Global, Local, Political* 425.

[9] Taedong Lee, *Global Cities and Climate Change: The Translocal Relations of Environmental Governance* (Routledge 2015).

[10] Sofie Bouteligier, *Cities, Networks and Global Environmental Governance: Spaces of Innovation, Places of Leadership* (Routledge 2013).

component of the contemporary international legal order.[11] She suggests that international law should not only concern itself with international organizations and the relationships amongst states. She also argues that the conception of the state as a unitary entity is an outdated one. By uncovering how sub-components of states form transboundary networks to address global governance issues, she argues that these inter-connections form a promising basis for increasing the scope and quality of international cooperation. While Slaughter's analysis of the disaggregated state provides a useful starting point for understanding and conceptualizing the governance activities of cities and their formation of transnational networks, it is, at the same time, 'a testament to the invisibility of cities to international lawyers'.[12] Slaughter describes judicial, legislative, and regulatory networks as examples of sub-state networks but does not discuss transnational city networks at all. Her vision of an international world order based on disaggregated states does not include the involvement of cities; yet, as I have sought to demonstrate, global cities have norm-setting governance capabilities and the potential to make a valuable contribution to the world order. This book therefore fills a gap in the literature on disaggregated states by examining the role of cities.

This research also makes a novel contribution to the voluntary standards literature, which has yet to address the standards and voluntary certification schemes that makes up urban climate law. There has been a steady increase in the use of voluntary standards to address environmental and social externalities across a range of economic sectors, including the global garment industry and food production.[13] This has prompted legal scholars to consider the regulatory and normative issues surrounding voluntary standards, such as their legitimacy and accountability, as well as the interaction between voluntary standards and 'hard law' legal instruments. This research expands the voluntary standards literature by shedding light on the voluntary standards and norms generated by cities and transmitted through their networks, which have hitherto not been studied from this perspective.

Finally, this book contributes to the emerging field of climate law. Climate change law, or climate law, has only very recently taken off as a distinct area of study. It has only been a matter of decades since lawyers first recognized the emergence of environmental law as a distinct field. Since then, environmental law has developed rapidly and now encompasses a number of sub-specializations, including water law and international environmental law.[14] As Jacqueline Peel puts it, '[t]he latest branch of the metaphorical environmental legal tree to take

[11] Anne-Marie Slaughter, *A New World Order* (Princeton University Press 2005).

[12] G. E. Frug and David Barron, 'International Local Government Law' (2006) 38 *Urban Lawyer* 1, pg. 23.

[13] See discussion in Chapter 6.

[14] See, for example, Philippe Cullet et al. (eds), *Water Law for the Twenty-First Century: National and International Aspects of Water Law Reform in India* (Routledge 2011); Philippe Sands and Jacqueline Peel, *Principles of International Environmental Law* (3rd edn, Cambridge University Press 2012).

shape is that of climate change law'.[15] Climate law can be defined as the distinctive body of legal principles and rules that have emerged because of the growing volume and complexity of regulatory activity around climate change at multiple levels and sites of governance.

Kati Kulovesi has identified two trends in climate law scholarship. The first is growing recognition that climate change has to be governed at multiple levels and not just at the international level. She argues that '[a]s a result, questions concerning the interplay between various sources of legal authority, including their hierarchies, synergies and tensions, are particularly relevant for climate law research and would arguably benefit from increased doctrinal attention'.[16] The second trend relates to the involvement of multiple non-state actors in climate change governance and the increasing reliance on soft law instruments and informal collaboration.[17] She argues that accounting for this plurality of actors and regulatory instruments involves some important challenges, 'including how to avoid becoming overtly descriptive and retain a normative focus'.[18]

This book contributes to the emerging field of climate law by providing a focused study on the norms, practices, and voluntary standards that have the potential to significantly alter GHG emissions patterns and climate adaptation practices at the local level. This research tries to uncover the unique normative contribution that cities are making towards efforts to address climate change, thereby adding to the increasingly rich fabric of climate law. Furthermore, this research takes heed of the implications of the trends in climate law scholarship that Kulovesi has identified and seeks to contribute to the existing climate law scholarship by providing a study that maintains a normative focus on the role of cities in governing climate change.

8.3.2 *Practical Implications*

This book offers policymakers and international organizations practical insight into the potential gains that can be reaped from working with city governments to implement urban climate law. The norms, practices, and voluntary standards that cities have developed and are implementing through their networks have the potential to lead to tonnes of GHG emissions reduction and to lay the foundation for a low-carbon future.

[15] Jacqueline Peel, 'Climate Change Law: The Emergence of a New Legal Discipline' (2008) 32 *Melbourne University Law Review* 922, pg. 923. For discussion about the relationship between environmental law and climate change law, see Chris Hilson, 'It's All about Climate Change, Stupid! Exploring the Relationship between Environmental Law and Climate Law' (2013) 25 *Journal of Environmental Law* 359.

[16] Kati Kulovesi, 'Exploring the Landscape of Climate Law and Scholarship: Two Emerging Trends' in Erkki J. Hollo, Kati Kulovesi, and Michael Mehling (eds), *Climate Change and the Law* (Springer 2013), pg. 32.

[17] Ibid. [18] Ibid., pg. 33.

Cities today are playing a role in addressing complex global issues such as climate change, as they are in implementing global human rights norms at the local level. Who is to say that cities and their governments will not also play a role in tackling human trafficking, promoting biological diversity, and combatting infectious diseases in the future? These are just a few of the global governance challenges we face today which require innovative and practical solutions. This book has shown that global cities are beginning to play a transnational lawmaking role in the area of climate change, suggesting the possibility that global cities can perform a similar role to address other global governance challenges. Those who are placed in charge of designing global regulatory frameworks in the future may wish to consider how to incorporate and leverage the governance potential of cities. Just as global cities have created C40 to scale up their climate actions and facilitate the transmission of best practices, they may be in a position to take similar actions for global health, for example.

8.4 AVENUES FOR FUTURE RESEARCH

Several avenues for further research present themselves. The research areas that are suggested next build upon this book and also fill gaps that have previously been identified. First, I have focused on cities, and specifically global cities, but cities are only one of many sub-national entities. Other sub-national entities include provinces, states within a federal system, and regions. These other sub-national governments could also play a role in transnational climate change governance, offering interesting possibilities for crossing the domestic-foreign divide and linking domestic climate change policies. For example, in a commentary on efforts by California (US) and the Brazilian state of Acre to link the latter's Reduction of Emissions from Deforestation and Forest Degradation (REDD) programme with the former's cap and trade emissions scheme, Ernesto Roessing Neto provides a fascinating insight into how sub-national action can provide a proof of concept that may influence the development of REDD at the international level and similar initiatives at the sub-national level in the future.[19] The role of sub-national actors in climate change law and policy nonetheless remains understudied and provides a fertile area for future research.

Secondly, there has been limited consideration in the literature about the legal capacity of cities and other sub-national actors for engaging in international relations. As Joana Setzer puts it, 'the governance literature has made limited strides with regard to the legal and institutional basis for international climate action by non-state actors. Rescaling processes are often taken for granted, with little or no consideration of whether the actors have a legal basis for moving across levels of

[19] Ernesto R. Neto, 'Linking Subnational Climate Change Policies: A Commentary on the California–Acre Process' (2015) 4 *Transnational Environmental Law* 425.

governance. For instance, are the representatives of subnational governments legally entitled to meet foreign dignitaries, to sign memoranda of understanding (MOUs) with other subnational governments across borders, or to establish emissions trading schemes?'[20] This book has not engaged with this question of legal competence, as urban climate law takes the form of voluntary certification and adoption of best practices, which do not require cities to exercise legal powers and enter into formal legal relations. However, Setzer raises an important point that in heralding the rise of cities as governance actors, further research is needed to understand the legal scope and limitations of sub-national climate action.

In recent times, much has been written about the potential fragmentation of the international legal system, the demise of the state as it loses its monopoly on authority and regulatory power, and the role of various non-state actors in international lawmaking and implementation. However, as I have mentioned earlier, very little consideration has been given to the opportunities and challenges posed to the modern state when the state disaggregates itself and sub-national entities emerge as actors in transnational governance. This book has shed light on how cities are emerging as norm-setting global governance actors in the context of climate change, but this is not the only area of international affairs that has witnessed the rise of cities. In a recently published collection, Barbara Oomen and her colleagues bring together academics and practitioners to consider the implications of the rise of 'human rights cities', a term that refers to cities that explicitly base their local policies on human rights.[21] Human rights cities forge alliances with international organizations and develop new practices designed to bring about 'global urban justice'.[22] It is clear that, at least in the areas of climate change and human rights, urban actors are becoming more prominent, and it can be argued that this trend will continue and reach other global policy areas. Meanwhile, international legal scholarship needs to keep apace with these developments and can make a valuable contribution towards shaping the normative discourse.

8.5 FINAL REMARKS

It is said that one chooses to view a glass of water as being half-empty or half-full. In the two decades since the UNFCCC came into force, there has been a significant increase in our collective knowledge about the causes and impacts of climate change, the crafting of creative solutions, and the development of a comprehensive international legal framework to galvanize climate action by states and other actors. While much remains to be done, and there are significant obstacles

[20] Joana Setzer, 'Testing the Boundaries of Subnational Diplomacy: The International Climate Action of Local and Regional Governments' [2015] 4 *Transnational Environmental Law* 319, pg. 326.

[21] Barbara Oomen, Martha F. Davis, and Michele Grigolo (eds), *Global Urban Justice: The Rise of Human Rights Cities* (Cambridge University Press 2016).

[22] Ibid.

that continue to hinder ambitious action by national governments, this book has shown that there is reason for optimism. Amidst relative inaction by national governments, cities have proactively become involved in the transnational climate change governance arena. Cities have not only taken action within their territories; they have developed best practices, voluntary standards, and norms that they seek to share and implement across borders. In the course of doing so, cities are reshaping traditional conceptions of international lawmaking and demonstrating the potential to play a meaningful role in other areas of global governance.

Select Bibliography

Abbott K. W., 'The Transnational Regime Complex for Climate Change' (2012) 30 *Environment and Planning C: Government and Policy* 571

'Strengthening the Transnational Regime Complex for Climate Change' (2014) 3 *Transnational Environmental Law* 57

Abbott K. W., Genschel P., Snidal D., and Zangl B. (eds), *International Organizations as Orchestrators* (Cambridge University Press 2015)

'Orchestration: Global Governance through Intermediaries' in Abbott K. W., Genschel P., Snidal D., and Zangl B. (eds), *International Organizations as Orchestrators* (Cambridge University Press 2015)

Abbott K. W. and Snidal D., 'Hard and Soft Law in International Governance' (2000) 54 *International Organization* 421

'The Governance Triangle: Regulatory Standards Institutions and the Shadow of the State' in Mattli W. and Woods N. (eds), *The Politics of Global Regulation* (Princeton University Press 2009)

'Strengthening International Regulation through Transnational New Governance: Overcoming the Orchestration Deficit' (2009) 42 *Vanderbilt Journal of Transnational Law* 501

Acuto M., 'Global Cities as Actors: A Rejoinder to Calder and de Freytas' (2009) 29 *SAIS Review of International Affairs* 175

'Global Cities: Gorillas in Our Midst' (2010) 35 *Alternatives: Global, Local, Political* 425

'Finding the Global City: An Analytical Journey through the "Invisible College"' (2011) 48 *Urban Studies* 2953

'City Leadership in Global Governance' (2013) 19 *Global Governance* 481

Global Cities, Governance and Diplomacy: The Urban Link (Routledge 2013)

'The New Climate Leaders?' (2013) 39 *Review of International Studies* 835

Adelman D. and Engel K., 'Adaptive Federalism: The Case against Reallocating Environmental Regulatory Authority' (2008) 92 *Minnesota Law Review* 1796

Adler J. H., 'Jurisdictional Mismatch in Environmental Federalism' (2005) 14 *New York University Environmental Law Journal* 130

Alger C. F., 'The World Relations of Cities: Closing the Gap between Social Science Paradigms and Everyday Human Experience' (1990) 34 *International Studies Quarterly* 493

'Expanding Governmental Diversity in Global Governance: Parliamentarians of States and Local Governments' (2010) 16 *Global Governance* 59

Alston P., 'The Myopia of the Handmaidens: International Lawyers and Globalization' (1997) 8 *European Journal of International Law* 435

Amin A. and Thrift N., 'Citizens of the World: Seeing the City as a Site of International Influence' (2005) 27 *Harvard International Review* 14

Andonova L. B., Betsill M. M., and Bulkeley H., 'Transnational Climate Governance' (2009) 9 *Global Environmental Politics* 52

Auld G. and Gulbrandsen L. H., 'Transparency in Nonstate Certification: Consequences for Accountability and Legitimacy' (2010) 10 *Global Environmental Politics* 97

Aust H. P., 'Shining Cities on the Hill? The Global City, Climate Change, and International Law' (2015) 26 *European Journal of International Law* 255

Avant D. D., Finnemore M., and Sell S. K. (eds), *Who Governs the Globe?* (Cambridge University Press 2010)

Barber B., *If Mayors Ruled the World: Dysfunctional Nations, Rising Cities* (Yale University Press 2014)

Berman P. S., 'A Pluralist Approach to International Law' (2007) 32 *Yale Journal of International Law* 301

Bernstein S., *The Compromise of Liberal Environmentalism* (Columbia University Press 2001)

Bernstein S. and Cashore B., 'Can Non-State Global Governance Be Legitimate? An Analytical Framework' (2007) 1 *Regulation & Governance* 347

Bernstein S. and Coleman W. D. (eds), *Unsettled Legitimacy: Political Community, Power, and Authority in a Global Era* (University of British Columbia Press 2009)

Betsill M. and Bulkeley H., 'Looking Back and Thinking Ahead: A Decade of Cities and Climate Change Research' (2007) 12 *Local Environment* 447

Betsill M. M. and Bulkeley H., 'Transnational Networks and Global Environmental Governance: The Cities for Climate Protection Program' (2004) 48 *International Studies Quarterly* 471

Biermann F., Pattberg P., van Asselt H., 'The Fragmentation of Global Governance Architectures: A Framework for Analysis' (2009) 9 *Global Environmental Politics* 14

Blank Y., 'Localism in the New Global Legal Order' (2006) 47 *Harvard International Law Journal* 263

Bodansky D., 'The United Nations Framework Convention on Climate Change: A Commentary' (1993) 18 *Yale Journal of International Law* 451

'A Tale of Two Architectures: The Once and Future U.N. Climate Change Regime' (2011) 43 *Arizona State Law Journal* 697

'Legitimacy in International Law and International Relations' in Dunoff J. L. and Pollack M. A. (eds), *Interdisciplinary Perspectives on International Law and International Relations* (Cambridge University Press 2012)

'The Legal Character of the Paris Agreement' (2016) *Review of European Community and International Environmental Law*

Bouteligier S., *Cities, Networks and Global Environmental Governance: Spaces of Innovation, Places of Leadership* (Routledge 2013)

Brunnee J. and Hey E., 'Transparency and International Environmental Institutions' in Bianchi A. and Peters A. (eds), *Transparency in International Law* (Cambridge University Press 2013)

Brunnee J. and Toope S. J., 'International Law and Constructivism: Elements of an Interactional Theory of International Law' (2000) 39 *Columbia Journal of Transnational Law* 19

Buchanan A. and Keohane R. O., 'The Legitimacy of Global Governance Institutions' (2006) 20 *Ethics & International Affairs* 405

Bulkeley H. and Betsill M., 'Rethinking Sustainable Cities: Multilevel Governance and the "Urban" Politics of Climate Change' (2005) 14 *Environmental Politics* 42

Bulkeley H., Andonova L., Bäckstrand K. et al., 'Governing Climate Change Transnationally: Assessing the Evidence from a Database of Sixty Initiatives' (2012) 30 *Environment and Planning C: Government and Policy* 591

Bulkeley H., Andonova L., Betsill M. et al., *Transnational Climate Change Governance* (Cambridge University Press 2014)

Bulkeley H. and Kern K., 'Local Government and the Governing of Climate Change in Germany and the UK' (2006) 43 *Urban Studies* 2237

Buthe T. and Mattli W., *The New Global Rulers: The Privatization of Regulation in the World Economy* (Princeton University Press 2011)

Caves, R. W. (ed), *Encyclopedia of the City* (Routledge 2005)

Chan, D. K.-H., 'City Diplomacy and "Glocal" Governance: Revitalizing Cosmopolitan Democracy' (2016) 29 *Innovation: The European Journal of Social Science Research*

Cheng, T. K., 'Convergence and Its Discontents: A Reconsideration of the Merits of Convergence of Global Competition Law' (2012) 12 *Chicago Journal of International Law* 433

Chimni, B. S., 'International Institutions Today: An Imperial Global State in the Making' (2004) 15 *European Journal of International Law* 1

Chinkin C., 'Normative Development in the International Legal System' in Shelton D. (ed), *Commitment and Compliance: The Role of Non-Binding Norms in the International Legal System* (Oxford University Press 2003)

Cho S. and Kelly C. R., 'Promises and Perils of New Global Governance: A Case of the G20' (2012) 12 *Chicago Journal of International Law* 491

Cole D. H., 'From Global to Polycentric Climate Governance' (2011) 2 *Climate Law* 395

Craig P. and de Búrca G., *The Evolution of EU Law* (Oxford University Press 2011)

de Burca G., Keohane R. O., and Sabel C., 'New Modes of Pluralist Global Governance' (2013) 45 *New York University Journal of International Law and Politics* 723

de Burca G. and Scott J., *Law and New Governance in the EU and the US* (Hart Publishing 2006)

den Exter R., Lenhart J., and Kern K., 'Governing Climate Change in Dutch Cities: Anchoring Local Climate Strategies in Organisation, Policy and Practical Implementation' (2014) *Local Environment: The International Journal of Justice and Sustainability* 1062

Depledge J. and Yamin F., 'The Global Climate-Change Regime: A Defence' in Helm D. and Hepburn C. (eds), *The Economics and Politics of Climate Change* (Oxford University Press 2009)

Dingwerth K. and Eichinger M., 'Tamed Transparency: How Information Disclosure under the Global Reporting Initiative Fails to Empower' (2010) 10 *Global Environmental Politics* 74

Elsig M., 'Orchestration on a Tight Leash: State Oversight of the WTO' in Abbott K. W., Genschel P., Snidal D., and Zangl B. (eds), *International Organizations as Orchestrators* (Cambridge University Press 2015)

Engel K. H. and Orbach B. Y., 'Micro-Motives and State and Local Climate Change Initiatives' 2 *Harvard Law & Policy Review* 119

Engel K. H. and Saleska S. R., 'Subglobal Regulation of the Global Commons: The Case of Climate Change' 32 *Ecology Law Quarterly* 183

Esty D., 'Revitalizing Environmental Federalism' (1996) 95 *Michigan Law Review* 570

Faure M., Smedt P. D., and Stas A. (eds), *Environmental Enforcement Networks: Concepts, Implementation and Effectiveness* (Edward Elgar Publishing 2015)
Fay C., 'Think Locally, Act Globally: Lessons to Learn from the Cities for Climate Protection Campaign' (2007) 7 *Innovations* 1
Feldman D. L., 'The Future of Environmental Networks-Governance and Civil Society in a Global Context' (2012) 44 *Futures* 787
Finnemore M. and Sikkink K., 'International Norm Dynamics and Political Change' (1992) 52 *International Organization* 887
Finnemore M. and Toope S. J., 'Alternatives to "Legalization": Richer Views of Law and Politics' (2001) 55 *International Organization* 743
Franck T. M., *The Power of Legitimacy among Nations* (Oxford University Press 1990)
Franck T. M., *Fairness in International Law and Institutions* (Oxford University Press 1998)
Fransen L. W. and Kolk A., 'Global Rule-Setting for Business: A Critical Analysis of Multi-Stakeholder Standards' (2007) 14 *Organization* 667
Freedman-Schnapp M., 'A Sustainable City for All' [2013] Progressive Policies for New York City in 2013 and Beyond
Freestone D. and Streck C. (eds), *Legal Aspects of Implementing the Kyoto Protocol Mechanisms: Making Kyoto Work* (Oxford University Press 2005)
Frug G. E., 'City as a Legal Concept' (1980) 93 *Harvard Law Review* 1057
Frug G. E. and Barron D., 'International Local Government Law' (2006) 38 *Urban Lawyer* 1
Green J., 'Private Standards in the Climate Regime: The Greenhouse Gas Protocol' (2010) 12 *Business and Politics Article* 3
Rethinking Private Authority: Agents and Entrepreneurs in Global Environmental Governance (Princeton University Press 2013)
Gupta A., 'Transparency under Scrutiny: Information Disclosure in Global Environmental Governance' (2008) 8 *Global Environmental Politics* 1
Gupta J., Van Der Leeuw K., and De Moel H., 'Climate Change: A "Glocal" Problem Requiring "Glocal" Action' (2007) 4 *Environmental Sciences* 139
Hakelberg L., 'Governance by Diffusion: Transnational Municipal Networks and the Spread of Local Climate Strategies in Europe' (2014) 14 *Global Environmental Politics* 107
Haynes J., 'Transnational Religious Actors and International Politics' (2001) 22 *Third World Quarterly* 143
Higgins R., *Problems and Process: International Law and How We Use It* (Oxford University Press 1994)
Hoffmann M., *Climate Governance at the Crossroads: Experimenting with a Global Response after Kyoto* (Oxford University Press 2011)
Jasanoff S. (ed), *States of Knowledge: The Co-Production of Science and the Social Order* (Routledge 2004)
Keck M. E. and Sikkink K., *Activists beyond Borders: Advocacy Networks in International Politics* (Cornell University Press 1998)
Keohane R. O. and Nye J. S., 'Transgovernmental Relations and International Organizations' (1974) 27 *World Politics* 39
Power and Interdependence: World Politics in Transition (TBS The Book Service Ltd 1977)
Keohane R. O. and Victor D. G., The Regime Complex for Climate Change (Harvard Project on International Climate Agreements Discussion Paper 10-33, 2010)
'The Regime Complex for Climate Change' (2011) 9 *Perspectives on Politics* 7
Kern K. and Bulkeley H., 'Cities, Europeanization and Multi-Level Governance: Governing Climate Change through Transnational Municipal Networks' (2009) 47 *Journal of Common Market Studies* 309

Kern K. and Mol A. P. J., 'Cities and Global Climate Governance: From Passive Implementers to Active Co-Decision-Makers' in Kaldor M. and Stiglitz J. E. (eds), *The Quest for Security: Protection without Protectionism and the Challenge of Global Governance* (Columbia University Press 2013)

Klabbers J., 'The Redundancy of Soft Law' (1996) 65 *Nordic Journal of International Law* 167

Koh H. H., 'The 1998 Frankel Lecture: Bringing International Law Home' (1998) 35 *Houston Law Review* 623

Kuh K. F., 'Capturing Individual Harms' (2011) 35 *Harvard Environmental Law Review* 155

Kulovesi K., 'Exploring the Landscape of Climate Law and Scholarship: Two Emerging Trends' in Hollo E. J., Kulovesi K., and Mehling M. (eds), *Climate Change and the Law* (Springer 2013)

Kulovesi K., Morgera E., and Muñoz M., 'Environmental Integration and the Multifaceted International Dimensions of EU Law: Unpacking the EU's 2009 Climate and Energy Package' (2011) 48 *Common Market Law Review* 829

Lee T., 'Global Cities and Transnational Climate Change Networks' (2013) 13 *Global Environmental Politics* 108

Global Cities and Climate Change: The Translocal Relations of Environmental Governance (Routledge 2015)

Lin J. and Streck C., 'Mobilising Finance for Climate Change Mitigation: Private Sector Involvement in International Carbon Finance Mechanisms' (2009) 10 *Melbourne Journal of International Law* 70

Ljungkvist K., *Global City 2.0: From Strategic Site to Global Actor* (Routledge 2016)

Meidinger E., 'The Administrative Law of Global Private-Public Regulation: The Case of Forestry' (2006) 17 *European Journal of International Law* 47

Naiki Y., 'Assessing Policy Reach: Japan's Chemical Policy Reform in Response to the EU's REACH Regulation' (2010) 22 *Journal of Environmental Law* 171

'Trade and Bioenergy: Explaining and Assessing the Regime Complex for Sustainable Bioenergy' (2016) 27 *European Journal of International Law* 129

Osofsky H. M., 'Multiscalar Governance and Climate Change: Reflections on the Role of States and Cities at Copenhagen' (2010) 25 *Maryland Journal of International Law* 64

Osofsky H. M., 'Suburban Climate Change Efforts: Possibilities for Small and Nimble Cities Participating in State, Regional, National and International Networks' (2012) 22 *Cornell Journal of Law and Public Policy* 395

Pattberg P. and Stripple J., 'Beyond the Public and Private Divide: Remapping Transnational Climate Governance in the 21st Century' (2008) 8 *International Environmental Agreements* 367

Pauwelyn J., 'Is It International Law or Not, and Does It Even Matter?' in Pauwelyn J., Wessel R. and Wouters J. (eds), *Informal International Lawmaking* (Oxford University Press 2012)

Pauwelyn J., Wessel R., and Wouters J., 'Informal International Lawmaking: An Assessment and Template to Keep It Both Effective and Accountable' in Pauwelyn J., Wessel R., and Wouters J. (eds), *Informal International Lawmaking* (Oxford University Press 2012)

Peel J., 'Climate Change Law: The Emergence of a New Legal Discipline' (2008) 32 *Melbourne University Law Review* 922

Porras I., 'The City and International Law: In Pursuit of Sustainable Development' (2008) 36 *Fordham Urban Law Journal* 537

Rajamani L., 'The Making and Unmaking of the Copenhagen Accord' (2010) 59 *International and Comparative Law Quarterly* 824

Ramsamy E., *World Bank and Urban Development: From Projects to Policy* (Routledge 2006)

Raustiala K., 'The Architecture of International Cooperation: Transgovernmental Networks and the Future of International Law' (2002) 43 *Virginia Journal of International Law* 1
Sassen S., *The Global City: New York, London, Tokyo* (Princeton University Press 1991)
Scott J., 'From Brussels with Love: The Transatlantic Travels of European Law and the Chemistry of Regulatory Attraction' (2009) 57 *American Journal of Comparative Law* 897
Scott J. and Rajamani L., 'EU Climate Change Unilateralism' (2012) 23 *European Journal of International Law* 469
Shaffer G. C., 'Transnational Legal Process and State Change' (2012) 37 *Law and Social Inquiry* 229
Shaffer G. C. and Bodansky D., 'Transnationalism, Unilateralism and International Law' (2012) 1 *Transnational Environmental Law* 31
Slaughter A.-M., *A New World Order* (Princeton University Press 2005)
Streck C. and Lin J., 'Making Markets Work: A Review of CDM Performance and the Need for Reform' (2008) 19 *European Journal of International Law* 409
van Asselt H., *The Fragmentation of Global Climate Governance: Consequences and Management of Regime Interactions* (Edward Elgar 2014)
Vogel D., *Trading Up: Consumer and Environmental Regulation in a Global Economy* (Harvard University Press 1995)
Wiener J. B., 'Think Globally, Act Globally: The Limits of Local Climate Policies' (2007) 155 *University of Pennsylvania Law Review* 1961
Yamin F. and Depledge J., *The International Climate Change Regime: A Guide to Rules, Institutions and Procedures* (Cambridge University Press 2004)

INTERVIEWS

1. Government official of the Hong Kong Special Administrative Region; interview conducted on 31 May 2016, Hong Kong
2. Staff member, Matchmaker Program; interview conducted by Skype on 22 June 2016
3. Project officer, Cities Development Initiative in Asia; interview conducted by Skype on 27 June 2016
4. Analyst, CDP; interview conducted by Skype on 28 June 2016
5. Manager, C40; interview conducted by Skype on 30 June 2016
6. Project leader, Centro Mario Molina; interview conducted by Skype on 12 July 2016
7. Vice-Mayor, mid-sized city in Greece; interview conducted on 5 July 2016 in Bonn, Germany
8. Project officer, Clean Tech Delta; interview conducted on 6 July 2016 in Rotterdam, the Netherlands
9. Government official, Mayor's Office of Recovery and Resiliency (New York City); interview conducted on the telephone on 5 July 2016
10. Program Manager, Hewlett Foundation; interview conducted by Skype on 8 July 2016
11. Government official, Rotterdam Municipal Government; interview conducted on 18 July 2016 in Rotterdam, the Netherlands
12. Program Officer, Greater London Authority; interview conducted via telephone on 21 July 2016
13. Government official, Regional Environmental Protection Agency (Rijnmond region); interview conducted on 25 July 2016 in Rotterdam, the Netherlands
14. Analyst, Bloomberg Associates; interview conducted via telephone on 8 August 2016

Index